Communications
in Computer and Information Science   1872

Rationale

The CCIS series is devoted to the publication of proceedings of computer science conferences. Its aim is to efficiently disseminate original research results in informatics in printed and electronic form. While the focus is on publication of peer-reviewed full papers presenting mature work, inclusion of reviewed short papers reporting on work in progress is welcome, too. Besides globally relevant meetings with internationally representative program committees guaranteeing a strict peer-reviewing and paper selection process, conferences run by societies or of high regional or national relevance are also considered for publication.

Topics

The topical scope of CCIS spans the entire spectrum of informatics ranging from foundational topics in the theory of computing to information and communications science and technology and a broad variety of interdisciplinary application fields.

Information for Volume Editors and Authors

Publication in CCIS is free of charge. No royalties are paid, however, we offer registered conference participants temporary free access to the online version of the conference proceedings on SpringerLink (http://link.springer.com) by means of an http referrer from the conference website and/or a number of complimentary printed copies, as specified in the official acceptance email of the event.

CCIS proceedings can be published in time for distribution at conferences or as postproceedings, and delivered in the form of printed books and/or electronically as USBs and/or e-content licenses for accessing proceedings at SpringerLink. Furthermore, CCIS proceedings are included in the CCIS electronic book series hosted in the SpringerLink digital library at http://link.springer.com/bookseries/7899. Conferences publishing in CCIS are allowed to use Online Conference Service (OCS) for managing the whole proceedings lifecycle (from submission and reviewing to preparing for publication) free of charge.

Publication process

The language of publication is exclusively English. Authors publishing in CCIS have to sign the Springer CCIS copyright transfer form, however, they are free to use their material published in CCIS for substantially changed, more elaborate subsequent publications elsewhere. For the preparation of the camera-ready papers/files, authors have to strictly adhere to the Springer CCIS Authors' Instructions and are strongly encouraged to use the CCIS LaTeX style files or templates.

Abstracting/Indexing

CCIS is abstracted/indexed in DBLP, Google Scholar, EI-Compendex, Mathematical Reviews, SCImago, Scopus. CCIS volumes are also submitted for the inclusion in ISI Proceedings.

How to start

To start the evaluation of your proposal for inclusion in the CCIS series, please send an e-mail to ccis@springer.com.

Gabriele Kotsis · A Min Tjoa · Ismail Khalil ·
Bernhard Moser · Atif Mashkoor ·
Johannes Sametinger · Maqbool Khan
Editors

Database and Expert Systems Applications - DEXA 2023 Workshops

34th International Conference, DEXA 2023
Penang, Malaysia, August 28–30, 2023
Proceedings

 Springer

Editors
Gabriele Kotsis
Johannes Kepler University Linz
Linz, Austria

Ismail Khalil
Johannes Kepler University Linz
Linz, Austria

Atif Mashkoor
Johannes Kepler University Linz
Linz, Austria

Maqbool Khan
Pak-Austria Fachhochschule - Institute
of Applied Sciences and Technology
(PAF-IAST)
Haripur, Pakistan

A Min Tjoa ⓘ
Vienna University of Technology
Vienna, Austria

Bernhard Moser
Software Competence Center Hagenberg
GmbH
Hagenberg, Austria

Johannes Sametinger
Johannes Kepler University Linz
Linz, Austria

ISSN 1865-0929   ISSN 1865-0937 (electronic)
Communications in Computer and Information Science
ISBN 978-3-031-39688-5   ISBN 978-3-031-39689-2 (eBook)
https://doi.org/10.1007/978-3-031-39689-2

This Springer imprint is published by the registered company Springer Nature Switzerland AG
The registered company address is: Gewerbestrasse 11, 6330 Cham, Switzerland

Preface

It is with great pleasure that we introduce the proceedings of DEXA 2023 workshops: the 7th International Workshop on Cyber-Security and Functional Safety in Cyber-Physical Systems (IWCFS 2023) and the 3rd International Workshop on AI System Engineering: Math, Modelling and Software (AISys 2023). These workshops serve as a global platform for researchers, engineers, and practitioners to explore the intricate realm of AI system engineering, with a specific focus on the essential aspects of mathematics, modelling, software development, cyber-security, and functional safety in the context of cyber-physical systems.

In recent years, the rapid advancement of artificial intelligence has permeated virtually every aspect of our lives, revolutionized industries and transformed the way we interact with the world. From autonomous vehicles and smart manufacturing to healthcare systems and smart cities, the integration of AI into cyber-physical systems has enabled unprecedented levels of efficiency, innovation, and automation. However, it also brings forth new challenges and complexities that demand rigorous engineering practices to ensure the reliability, security, and safety of these systems.

The papers featured in this volume represent the cutting-edge research, insights, and innovations shared during the conference. They cover a wide range of topics, including mathematical foundations for AI, modelling techniques, software engineering methodologies, cyber-security considerations, and functional safety approaches in the development and deployment of AI systems within the realm of cyber-physical systems.

Each paper included in this compilation has undergone a rigorous, at least 3 reviewers per submission in a single-blind review process, ensuring the highest standards of quality and relevance to the conference theme. The expertise and dedication of the authors, combined with the invaluable feedback from the program committee, have contributed to the depth and breadth of knowledge presented in this volume. Out of 20 submissions, 7 were accepted as full papers and 3 were accepted as short papers.

We would like to extend our sincere appreciation to all the authors for their valuable contributions to this collection. Their tireless efforts and pioneering research have significantly advanced the field of AI system engineering, bringing us closer to the realization of safe, secure, and reliable AI-driven cyber-physical systems. We would also like to express our gratitude to the program committee members for their expertise, rigor, and dedication in evaluating and selecting the papers.

Furthermore, we want to extend our appreciation to the conference organizers, sponsors, and participants for their unwavering support and enthusiastic engagement. It is through their collective efforts that these workshops have become a thriving platform for knowledge exchange, interdisciplinary collaboration, and innovation in the field of AI system engineering.

As you explore the pages of these proceedings, we hope that you find inspiration in the pioneering research, novel methodologies, and insightful discussions presented here. Whether you are a researcher, an engineer, a practitioner, or a policy-maker, we believe that the wealth of knowledge shared within this volume will not only expand your understanding but also inspire you to embrace the challenges and opportunities that lie ahead in AI system engineering for cyber-physical systems.

Finally, we would like to express our heartfelt appreciation to the entire community. It is your collective dedication, expertise, and commitment to advancing the field of AI system engineering that made this event a resounding success.

August 2023
Gabriele Kotsis
A. Min Tjoa
Ismail Khalil
Bernhard Moser
Atif Mashkoor
Johannes Sametinger
Maqbool Khan

Organization

Steering Committee

Gabriele Kotsis	Johannes Kepler University Linz, Austria
A Min Tjoa	Vienna University of Technology, Austria
Robert Wille	Software Competence Center Hagenberg, Austria
Bernhard Moser	Software Competence Center Hagenberg, Austria
Ismail Khalil	Johannes Kepler University Linz, Austria

AISys 2023 Chairs

Paolo Meloni	University of Cagliari, Italy
Maqbool Khan	PAF-IAST, Pakistan
Gerald Czech	Upper Austrian Fire Brigade Association, Austria
Thomas Hoch	Software Competence Center Hagenberg, Austria
Bernhard Moser	Software Competence Center Hagenberg, Austria

AISys 2023 Program Committee

Paola Busia	University of Cagliari, Italy
Sajid Shah	Prince Sultan University, Saudi Arabia
Nazeer Muhammad	PAF-IAST, Pakistan
Rizwan Ullah	Chulalongkorn University, Thailand
Xiaolong Xu	Nanjing University of Information Science and Technology, China

IWCFS 2023 Chairs

Atif Mashkoor	LIT Secure & Correct Systems Lab, Austria
Johannes Sametinger	Johannes Kepler University Linz, Austria

IWCFS 2023 Program Committee

Yamine Ait Ameur	INPT-ENSEEIHT/IRIT, France
Paolo Arcaini	National Institute of Informatics, Japan

Richard Banach	University of Manchester, UK
Ladjel Bellatreche	ENSMA, France
Silvia Bonfanti	University of Bergamo, Italy
Jorge Cuellar	University of Passau, Germany
Angelo Gargantini	University of Bergamo, Italy
Irum Inayat	National University of Computer and Emerging Sciences, Pakistan
Jean-Pierre Jacquot	University of Lorraine, France
Muhammad Khan	University of Greenwich, UK
Saif Ur Rehman Khan	COMSATS University Islamabad, Islamabad Campus, Pakistan
Rudolf Ramler	Software Competence Center Hagenberg GmbH, Austria
Neeraj Singh	INPT-ENSEEIHT/IRIT, France
Edgar Weippl	University of Vienna, Austria

Organizers

Contents

AI System Engineering: Math, Modelling and Software

DSD: The Data Source Description Vocabulary

Lisa Ehrlinger[2]([✉])[iD], Johannes Schrott[1,2]([✉])[iD], and Wolfram Wöß[1]

[1] Johannes Kepler University Linz, Linz, Austria
{johannes.schrott,wolfram.woess}@jku.at
[2] Software Competence Center Hagenberg GmbH, Hagenberg, Austria
{lisa.ehrlinger,johannes.schrott}@scch.at

Abstract. Training machine learning models, especially in producing enterprises with numerous information systems having different data structures, requires efficient data access. Hence, standardized descriptions of data sources and their data structures are a fundamental requirement. We therefore introduce version 4.0 of the Data Source Description Vocabulary (DSD), which represents a data source in a standardized form using an ontology. We present several real-world applications where the DSD vocabulary has been applied in recent years to demonstrate its relevance. An evaluation against the FAIR principles highlights the scientific quality and potential for reuse of the DSD vocabulary.

Keywords: Data source representation · FAIR · Vocabulary · Ontology

1 Introduction

Training machine learning (ML) models [8], integration of heterogeneous data sources [5], or data quality measurement [3,4] are exemplary tasks that involve more than one data source in an organization. To merge these data sources, a standardized description of the data sources and their data structures is required. Data Source Description Vocabulary (DSD)[1] version 4.0, which enables the standardized representation of data sources and their internal structure independently of the original type of source (e.g., database management system, comma-separated values (CSV) files).

We delimit DSD from related research in Sect. 2 and describe the details of the vocabulary in Sect. 3. Sect. 4 highlights the relevance of DSD by outlining its applications in practice. The vocabulary is evaluated against the FAIR (Findability, Accessibility, Interoperability, and Reuse [12]) principles in Sect. 5.

2 Related Work

The idea of developing a standardized representation for data sources of different types is not new. Atzeni et al. [1] present a metamodel that can represent

---

[1] Available online: IRI: https://w3id.org/dsd; DOI: https://doi.org/10.5281/zenodo.7773861.

© The Author(s), under exclusive license to Springer Nature Switzerland AG 2023
G. Kotsis et al. (Eds.): DEXA 2023 Workshops, CCIS 1872, pp. 3–10, 2023.
https://doi.org/10.1007/978-3-031-39689-2_1

(amongst others) relational data models, Entity-Relationship models, and object-oriented models. Candel et al. [2] propose "U-Schema", a unified metamodel that is based on the Eclipse Modeling Framework (EMF)[2] and supports the most-widely used NoSQL systems, as well as MySQL. The DSD vocabulary is different from such metamodels since it is based on the Ontology Language (OWL)[3] for building ontologies that represent data sources.

The following OWL-based vocabularies for describing the metadata of data sources [13] have been recommended by the World WideWeb Consortium (W3C):

- the Data Catalog Vocabulary (DCAT)[4], which provides terms for describing so-called "data sets" (i.e., data sources) and services to catalog them, and
- the Vocabulary of Interlinked Datasets (VoID)[5], which is specifically tailored to describe metadata of Resource Description Framework (RDF) data sets.

In contrast to DSD, both vocabularies do not cover the structure inside a data source. There are also some vocabularies that support the representation of the internal structure of a data source, like CSV on the Web (CSVW)[6] that allows describing the structure of CSV files, or the RDF Data Cube Vocabulary[7] that is suitable for multidimensional data. All of these vocabularies are dedicated to a specific data source type, while DSD is data source type independent. The Semantic Data Dictionary (SDD) has a similar objective as DSD, but only supports tabular data in its current state (Extensible Markup Language (XML) is planned in the future) [10].

Despite the same acronym, the DSD vocabulary is also different from the DSD Schema Language [9], which is an XML schema language with higher expressiveness than the XML document type declaration (DTD)[8] or XML Schema (XSD)[9].

In summary, there is no other OWL-based vocabulary than DSD that can represent data sources, independently of their type and internal structure.

3   The Data Source Description Vocabulary (DSD)

Originally, Ehrlinger and Wöß published DSD in 2015 [5]. The vocabulary is based on OWL, RDF, and RDF Schema. The core idea of DSD is to provide a terminology for representing the structure of data sources independently of their type [5]. It can be used to represent different types of data sources (e.g., relational or graph databases, document stores) and their (internal) semantics.

Based on our experience in data modeling (Entity-Relationship (ER) models, Unified Modeling Language (UML), and ontologies) and on requirements raised

---

[2] https://www.eclipse.org/modeling/emf/.
[3] https://www.w3.org/TR/owl2-overview/.
[4] http://www.w3.org/ns/dcat#.
[5] http://rdfs.org/ns/void#.
[6] http://www.w3.org/ns/csvw#.
[7] http://purl.org/linked-data/cube#.
[8] https://www.w3.org/TR/REC-xml/#dt-doctype.
[9] https://www.w3.org/TR/xmlschema-0/.

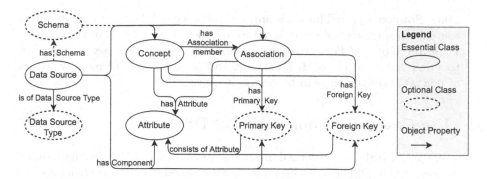

Fig. 1. OWL classes and OWL object properties in the DSD vocabulary

by company partners (cf. applications of DSD in Sect. 4), we defined a set of terms (i.e., OWL classes, object properties, and data properties) for describing data sources. Figure 1 illustrates the classes and object properties defined in DSD. For simplicity, inverse object properties are not shown. An inverse object property in OWL is a relationship between two classes where the direction of the relationship is reversed. We distinguish between "essential" classes, which are necessary for describing a data source using DSD, and "optional" classes, which provide additional non-necessary features. Below, we describe each class, in order of importance.

Essentials

- **Data Source.** A generic class for representing data sources. *Example:* A dsd:DataSource can represent structured data such as relational databases, semi-structured data like XML files, or NoSQL databases such as graph databases or wide-column stores.
- **Concept.** A representation of a structural part of a data source. *Example:* A dsd:Concept can represent a table or a view of a relational database or a class in object-oriented structures.
- **Attribute.** A dsd:Attribute describes a property of a dsd:Concept. DSD also provides OWL data properties to define certain attribute characteristics, such as, nullable or unique. *Example:* If a dsd:Concept represents a relational table, its attributes correspond to the columns.
- **Association.** A dsd:Association describes a relationship between two instances of dsd:Concept. There are three disjoint dsd:Association sub-classes for aggregation, inheritance, and reference associations. For further details and also for object properties of the subclasses, we refer to [5].

Optionals

- **Schema.** Instances of dsd:Schema create an optional hierarchy level between data sources (instances of dsd:DataSource) and concepts (instances of dsd:Concept). Schemas allow the grouping of concepts and are commonly used in enterprise databases.

- **Data Source Type.** This class provides instances of the most common data source types, which can be assigned to instance of dsd:DataSource.
- **Primary Key** and **Foreign Key.** Instances of these two classes are assigned to a dsd:Association or dsd:Concept and consist of one or more instances of dsd:Attribute (i.e., can be composite keys).

4   Use Cases and Applications of DSD

In recent years, DSD has been used in various applications. This section discusses three areas where DSD can be useful for both researchers and practitioners.

Schema Matching and Schema Similarity. A key advantage of DSD is to make data sources and their schemas comparable. Thus, in [6], DSD was used to generate homogeneous representations of data source schemas, which could then be compared directly. The similarity of these schemas (i.e., their degree of overlap) was used as input for a metric to assess the schema quality [6].

Metadata Management. The implementation of a corporate metadata management system (e.g., a data catalog) requires comparability of data source schemas from different types. For that purpose, we employed DSD to represent different data sources in a producing company [11]. In this project, DSD was the basis to describe data sources and their internal structure, which can then be annotated with different kinds of metadata, e.g., access security metadata or the assignment of data responsibility roles.

Data Quality. In real-world scenarios, data quality assessment should be carried out on multiple (heterogeneous) data sources. Thus, the data quality tools QuaIIe [4] and DQ-MeeRKat [3], which aim to be data source type independent, implement connectors[10] that map the original schema of a data source to a DSD representation (see Table 1 in [5]). After calculating different data quality metrics, the measurement results can be annotated to these representations.

5   Evaluation Against the FAIR Principles

The FAIR principles define a measurable set of guidelines to assess the FAIRness of a data asset [12] and are therefore well suited to evaluate the quality (i.e., findability, accessibility, interoperability, and reuse) of DSD. We conducted a twofold evaluation: (1) an automated evaluation using FOOPS![11] in Sect. 5.1 and (2) a manual evaluation with the FAIR principles published online in Sect. 5.2.

---

[10] See the "connectors" Java package in https://github.com/lisehr/dq-meerkat.
[11] https://w3id.org/foops/.

5.1   Automatic Evaluation

For the automatic evaluation, we used the tool FOOPS! (Ontology Pitfall Scanner for FAIR) [7]. FOOPS! determines FAIRness by checking if Internationalized Resource Identifiers (IRIs) are resolvable and permanent, and if certain OWL properties (e.g., author, publication date, provenance information) are present.

In the automatic evaluation, DSD achieves a FAIRness score of 88%. FOOPS! does not assess DSD to be fully FAIR since it does not recognize some specific metadata. As an example, information on authors and contributors of DSD is included as instances of `foaf:Person`, but FOOPS! expects the presence of literal values.

5.2   Manual Evaluation

For each FAIR principle[12], we manually assessed and justified if it is fulfilled by DSD, as shown in detail in Table 1. Overall, we consider DSD to be fully FAIR.

Table 1. Manual evaluation against the FAIR Principles.

FAIR principle	Fulfillment	Justification
Findable		
F1. (Meta)data are assigned a globally unique and persistent identifier.	✓	The base IRI of DSD is https://w3id.org/dsd, which is unique and a persistent identifier.
F2. Data are described with rich metadata (defined by R1 below)	✓	*See detailed principles R1.1-1.3.*
F3. Metadata clearly and explicitly include the identifier of the data they describe	✓	The metadata of the vocabulary is annotated using RDF. Data (= subject) is annotated with specific (= predicate) metadata (= object).
F4. (Meta)data are registered or indexed in a searchable resource	✓	DSD is indexed in Linked Open Vocabularies (LOV)[a].
Accessible		
A1. (Meta)data are retrievable by their identifier using a standardised communications protocol	✓	The vocabulary is available online (see Footnote 1) and can be retrieved using the HTTPS protocol.
A1.1 The protocol is open, free, and universally implementable	✓	HTTPS fulfills all these criteria.
A1.2 The protocol allows for an authentication and authorisation procedure, where necessary	✓	HTTPS allows, e.g., basic-auth. In the case of DSD, no authentication and authorization are needed.

(continued)

---

[12] The FAIR principles and the corresponding descriptions in the leftmost column of Table 1 are directly taken from the GO-FAIR website (https://www.go-fair.org/fair-principles/).

Table 1. (*continued*)

FAIR principle	Fulfillment	Justification
A2. Metadata are accessible, even when the data are no longer available	✓	DSD has a DOI and is indexed in LOV[a] as well as prefix.cc[b]. Furthermore, a GitHub repository[c] exists.
Interoperable		
I1. (Meta)data use a formal, accessible, shared, and broadly applicable language for knowledge representation	✓	DSD is available online in Turtle[d] syntax.
I2. (Meta)data use vocabularies that follow FAIR principles	✓	DSD is based on RDF and OWL. It does not import any other vocabularies.
I3. (Meta)data include qualified references to other (meta)data	✓	The metadata of DSD is encoded using RDF, thus all references are qualified.
Reusable		
R1. (Meta)data are richly described with a plurality of accurate and relevant attributes	✓	*See detailed principles R1.1-1.3*
R1.1. (Meta)data are released with a clear and accessible data usage license	✓	DSD is licensed under the GNU Lesser General Public Licens (LGPL)[e].
R1.2. (Meta)data are associated with detailed provenance	✓	Provenance information is provided via DSDs GitHub repository[c]. To maintain a clear scope of the vocabulary, we do not include provenance information directly in the vocabulary.
R1.3. (Meta)data meet domain-relevant community standards	✓	The vocabulary uses RDF and OWL. Metadata information of the vocabulary is encoded with terms that are recommended as "best-practice" by FOOPS! and PyLODE[f].

[a] https://lov.linkeddata.es/dataset/lov/
[b] https://prefix.cc/
[c] https://github.com/FAW-JKU/dsd-vocabulary
[d] https://www.w3.org/TR/2014/REC-turtle-20140225/
[e] https://www.gnu.org/licenses/old-licenses/lgpl-2.1.html
[f] https://github.com/RDFLib/pyLODE

6   Conclusion and Outlook on Future Work

Although the focus of DSD is on the description of data sources, previous versions contained, e.g., a class `Stakeholder`, which was used for modelling people and their permissions to data sources. In the newest version 4.0, we removed all capabilities that do not support the core idea of DSD and suggest the reuse and combination with other vocabularies to annotate different kinds of *metadata* to a data source. An example is the Data Quality Vocabulary (DQV)[13], which is specifically designed to represent data quality metadata. DSD 4.0 is the first version that includes a rich set of metadata as well as a permanent identifier, and thus fulfills the FAIR principles. Due to intensively using DSD in data quality tools (cf. [3,4]), we will further investigate the integration of DSD with DQV in our ongoing research. At this point, we would like to encourage other research groups to investigate the integration of additional vocabularies for annotating *metadata* to DSD data sources, e.g., security or provenance metadata.

All links in this publication were last visited on June 1, 2023.

Acknowledgements. This research has been partially funded by BMK, BMAW, and the State of Upper Austria in the frame of the SCCH competence center INTEGRATE (FFG grant no. 892418) part of the FFG COMET Competence Centers for Excellent Technologies Programme and by the "ICT of the Future" project QuanTD (no. 898626).

References

1. Atzeni, P., Gianforme, G., Cappellari, P.: A universal metamodel and its dictionary. In: Hameurlain, A., Küng, J., Wagner, R. (eds.) Transactions on Large-Scale Data- and Knowledge-Centered Systems I. LNCS, vol. 5740, pp. 38–62. Springer, Heidelberg (2009). https://doi.org/10.1007/978-3-642-03722-1_2
2. Candel, C.J.F., Sevilla Ruiz, D., García-Molina, J.J.: A Unified Metamodel for NoSQL and Relational Databases. Information Syst. **104**, 101898 (2022). https://doi.org/10.1016/j.is.2021.101898
3. Ehrlinger, L., Gindlhumer, A., Huber, L., Wöß, W.: DQ-MeeRKat: automating Data Quality Monitoring with a Reference-Data-Profile-Annotated Knowledge Graph. In: Proceedings of the 10th International Conference on Data Science, Technology and Applications - DATA, pp. 215–222. SciTePress (2021)
4. Ehrlinger, L., Werth, B., Wöß, W.: Automated continuous data quality measurement with QuaIIe. Int. J. Adv. Softw. **11**(3 & 4), 400–417 (2018)
5. Ehrlinger, L., Wöß, W.: Semi-automatically generated hybrid ontologies for information integration. In: Joint Proceedings of the Posters and Demos Track of 11th International Conference on Semantic Systems - SEMANTiCS2015 and 1st Workshop on Data Science: Methods, Technology and Applications (DSci15), vol. 1481, pp. 100–104. CEUR Workshop Proceedings (2015). https://ceur-ws.org/Vol-1481/paper30.pdf
6. Ehrlinger, L., Wöß, W.: Automated schema quality measurement in large-scale information systems. In: Hacid, H., Sheng, Q.Z., Yoshida, T., Sarkheyli, A., Zhou, R. (eds.) QUAT 2018. LNCS, vol. 11235, pp. 16–31. Springer, Cham (2019). https://doi.org/10.1007/978-3-030-19143-6_2
7. Garijo, D., Corcho, O., Poveda-Villalón, M.: FOOPS!: an ontology pitfall scanner for the FAIR principles. In: International Semantic Web Conference (ISWC) 2021. CEUR Workshop Proceedings, vol. 2980 (2021). http://ceur-ws.org/Vol-2980/paper321.pdf

---

[13] http://www.w3.org/ns/dqv#.

8. Gebru, T.: Datasheets for datasets. Commun. ACM **64**(12), 86–92 (2021). https://doi.org/10.1145/3458723
9. Klarlund, N., Møller, A., Schwartzbach, M.I.: The DSD schema language. Autom. Softw. Eng. **9**, 285–319 (2002). https://doi.org/10.1023/A:1016376608070
10. Rashid, S.M., et al.: The semantic data dictionary - an approach for describing and annotating data. Data Intell. **2**(4), 443–486 (2020). https://doi.org/10.1162/dint_a_00058
11. Schrott, J., Weidinger, S., Tiefengrabner, M., Lettner, C., Wöß, W., Ehrlinger, L.: GOLDCASE: a generic ontology layer for data catalog semantics. In: Garoufallou, E., Vlachidis, A. (eds.) MTSR 2022. CCIS, vol. 1789, pp. 26–38. Springer, Cham (2023). https://doi.org/10.1007/978-3-031-39141-5_3
12. Wilkinson, M.D., et al.: The FAIR Guiding Principles for scientific data management and stewardship. Sci. Data **3**(1), 160018 (2016). https://doi.org/10.1038/sdata.2016.18
13. World Wide Web Consortium: All Standards and Drafts - W3C. https://www.w3.org/TR/. Accessed 21 Feb 2023

Analyzing the Innovative Potential of Texts Generated by Large Language Models: An Empirical Evaluation

Oliver Krauss[1,2(✉)] [ID], Michaela Jungwirth[1,2], Marius Elflein[1,2],
Simone Sandler[1,2] [ID], Christian Altenhofer[1,2], and Andreas Stoeckl[1,3] [ID]

[1] University of Applied Sciences Upper Austria, Campus Hagenberg,
Softwarepark 11, 4232 Hagenberg, Austria
{oliver.krauss,simone.sandler,andreas.stoeckl}@fh-hagenberg.at
[2] Advanced Information Systems and Technology, Hagenberg, Austria
[3] Digital Media Department, Hagenberg, Austria
http://aist.science

Abstract. As large language models (LLMs) revolutionize natural language processing tasks, it remains uncertain whether the text they generate can be perceived as innovative by human readers. This question holds significant implications for innovation management, where the generation of novel ideas from extensive text corpora is crucial. In this study, we conduct an empirical evaluation of 2170 generated idea texts, containing product and service ideas in current trends for specific companies, focusing on three key metrics: innovativeness, context, and text quality. Our findings show that, while not universally applicable, a substantial number of LLM-generated ideas exhibit a degree of innovativeness. Remarkably, only 97 texts within the entire corpus were identified as highly innovative. Moving forward, an automated evaluation and filtering system to assess innovativeness could greatly support innovation management by facilitating the pre-selection of generated ideas.

Keywords: Artificial Intelligence · Decision Support · Large Language Models · Data Quality

1 Introduction

We examine the perception of innovativeness in texts generated by Large Language Models (LLMs) among a human audience. To address this question, we perform an empirical evaluation on a comprehensive set of 2170 texts generated specifically for innovation management. Our evaluation focuses on assessing the degree of *innovativeness*, the *context*ual relevance of the innovations proposed, and the extent to which the generated *text*s resemble human-written content.

It is widely recognized that Large Language Models (LLMs) lack thought or reasoning capabilities and can be likened to "Stochastic Parrots" [2]. These models rely on predicting the next words in a sequence based on vast corpora of data, without true comprehension. However, recent advancements have integrated LLMs with online source access, enabling them to retrieve information [9].

G. Kotsis et al. (Eds.): DEXA 2023 Workshops, CCIS 1872, pp. 11–22, 2023.
https://doi.org/10.1007/978-3-031-39689-2_2

The integration of next word prediction with real-time data access has shown promising results in simulating novelty within various domains. This approach has been particularly valuable in innovation management, facilitating data analysis and idea generation [3]. Furthermore, it has proven effective in expanding, combining, and explaining novel ideas through creative processes [8].

Previous research has delved into the novelty of texts generated by LLMs, with a specific focus on programming exercise generation [16]. Although an empirical evaluation was not conducted, the study aimed to determine if the generated texts could be found on the internet. Remarkably, the findings revealed that 81.8% of all generated examples were deemed novel. While novelty does not directly imply innovativeness, these results indicate the ability of LLMs to produce new information. In smaller models such as GPT-2, instances of text duplication may occur, diminishing novelty and innovation [13].

Our objective is to determine the true innovativeness of texts generated by LLMs. Our evaluation encompassed a diverse range of domains, including new ideas, products, improvements, and novel customer service areas for a total of 31 companies. We aim to answer the following questions:

- Perception of Innovativeness: We explore whether ideas generated by LLMs are perceived as truly *innovative* by human evaluators. This investigation delves into the subjective assessment of the generated texts in terms of their novelty and creativity.
- Contextual Relevance: We examine the degree to which the generated ideas align with the *context* of the specific company they were generated for. This analysis assesses the appropriateness of the ideas in relation to the company's industry, objectives, and current trends.
- Human-Like Text Quality: We evaluate the *text* quality of LLM-generated passages and investigate whether they can convincingly pass as human-written content. This assessment takes into account factors such as coherence, grammar, and overall fluency.

2   Background

LLMs are large models trained on general data to solve a wide range of Natural Language Processing (NLP) tasks. In order to perform the desired task, a prompt is given to the LLM [14]. Examples for LLMs are GPT3 [4], which was used to generate the text corpus, BLOOM [17] and PaLM [5]. LLMs are able to achieve state-of-the-art results on various NLP tasks, including text generation [15].

When evaluating the innovativeness of texts generated by LLMs, it is crucial to recognize the specific task to which the text corpus was applied: ideation in the domain of innovation management. After the trend scouting phase, ideation is used to generate and subsequently filter ideas based on their potential. This stage serves as a foundation for further evaluation and analysis, allowing more promising ideas to be explored while discarding less viable ones [6].

The primary concept underlying the text corpus employed in our research was to leverage Language Models (LLMs) for analyzing extensive collections of data and generating fresh ideas. This approach aimed to provide innovation managers with one concise paragraph of text per idea that serve as valuable starting points. Manual sifting through such corpora would pose a significant challenge for humans; however, the utilization of LLMs holds promising potential for enhancing innovation management [7].

In our case the trends were mined from news articles, scientific publications from conferences and journals as well as arXiv documents. The trends were mined via topic modelling [10], of the title and abstract of each article.

We employed GPT-3, to create an idea that may be usable by a business to expand on their current portfolio, or to create new products or services. Our aim was to harness the capabilities of GPT-3 to generate ideas that businesses could potentially utilize to expand their existing portfolio or develop new products and services. The paramount goal of these generated ideas was to appear genuinely *innovative* to the readers. Additionally, the business provided us with a specific *context* to determine the innovation potential of the ideas. Furthermore, the LLM was equipped with knowledge of a specific trend, backed by relevant articles, to enhance its understanding of the current market landscape.

3 Methods

We present the experiment design, which involved conducting an empirical analysis of 2170 texts generated by GPT-3. Each of these texts corresponded to a unique idea generated based on a current industry trend for a specific company. To assess the quality and effectiveness of these ideas, we enlisted a group of 11 reviewers who evaluated text across 3 key metrics:*innovative*, *context*, and *text*. Each text underwent evaluation by 3 out of 11 reviewers to ensure a comprehensive and robust assessment process.

We selected 31 Austrian companies representing various industries, such as *IT*, *Law* or *Energy*. For each of these companies, we identified 10 current industry trends for which ideas were to be generated. Subsequently, we generated 7 distinct ideas for each of the identified trends. This results in a total of 2170 texts (31 companies x 10 trends x 7 ideas). This comprehensive approach ensured a diverse and extensive text corpus for our analysis.

For each of the 31 companies, a profile was created which has the following information, which is also presented in Table 1:

name of the company. The actual names of the companies were replaced with single words that broadly represent their core industry, ensuring anonymity throughout the work.

website of the company. Provided to the reviewers and not GPT-3, to assess if the generated text is fitting the context of the company.

keywords consisting of 5 keywords (e.g. "food") or keyword phrases (e.g. "food science"), representing the core industrial areas or products.

description summarizing a company was provided to the reviewers and GPT-3.

Table 1. Example profile of the company *Food* from which reviewers can assess the context of the idea.

Data	Example
name	*Food*
website	https://www.example-food.com/
keywords	food science, food processing, fruit, starch, sugar
description	*Food* is an Austrian company which produces a wide range of industrial products for the processing sector. In its business segments, Fruit, Starch and Sugar, *Food* supplies local producers and large international players, particularly those in the food processing industry

For each of the trends, a description (see Table 2) was formulated and was made available to GPT-3. Similarly, the reviewers had access to this description to assess the alignment of the generated idea texts with both the trend and the company's context. The trends were extracted from a collection of articles, and the trend titles and keywords were derived from the entire set of articles. The trend description incorporated the titles and summaries of the top 10 articles pertaining to that specific trend.

To ensure the integrity and quality of the generated text and maintain consistency in the evaluation process, certain information from the articles was intentionally excluded. This includes the source or news outlet and the full text of the articles. Both GPT-3 and the reviewers were provided with short summaries of the articles. By providing only summaries of the articles to GPT-3, the potential issue of verbatim copy-pasting from the articles was mitigated, which has been identified as a challenge in text generation using Language Models (LLMs) [13]. The trend descriptions included the following details:

title of the trend
keywords of the trend, mined from the entire article set available for a trend. The reviewers were presented with the top 5 words.
articles comprising summaries of the top 10 articles. Each article consisted of the following information:
    title short title of the article.
    description containing a short summary of the article.

The corpus of 2170 texts, derived from the company and trend descriptions, underwent analysis by a total of 11 reviewers. Each text was evaluated by 3 reviewers, who assigned individual scores for the following metrics. All metrics were assessed on a scale of 0 to 5:

innovative 0 indicates a lack of any innovative elements and may come across as generic or unoriginal. 5 suggests a novel and highly specific idea.
context 0 signifies that the text fails to align with either the trend or the company's context. 5 fits well within the given trend and the company.
text 0 implies that the text exhibits clear faults or incoherence, 5 is easily readable, and could convincingly pass as human-written.

Table 2. Example description of a trend for company *Food*. The articles 3-10 are not shown.

Data	Example
title	Plant-based sugar substitute
keywords	fruit quality, sugar content, ingredients, nutritional, fruit weight
article 1 title	Chocolate food product
article 1 summary	A chocolate food product that is low in fat, dairy free, soy and lecithin free, free of added sugar or ingredients that increase sugar content, substantially starch-free, Isomaltooligosaccharide-free, oligosaccharide-free, maltitol-free, sorbitol-free, xylitol-free, erythritol-free, and isomalt-free. The chocolate food product may comprise: a cocoa butter, an unsweetened cocoa powder, a glycerin, a coconut cream, an almond milk, a pectin, a salt, a monk fruit blend, and a coconut flour.
article 2 title	Agent for increasing sugar content in fruit
article 2 summary	Provided is a compound and composition capable of increasing a sugar content in a fruit by a simple method, without being restricted by a cultivation area of a plant or a climatic environment. The agent for increasing a sugar content in a fruit of a plant comprises a compound represented by the formula MX as an active ingredient, wherein M represents alkali metal ion or alkaline earth metal ion, and X represents carbonate ion, hydrogen carbonate ion, acetate ion, citrate ion, succinate ion, phosphate ion, hydrogen phosphate ion, or pyrophosphate ion

The setup involved a systematic process for evaluating the texts. A reviewer selected a company and was provided with the company profile, outlined in Table 1. The reviewer received the 10 trends associated with the company, and was required to read through all 10 short summary articles for each trend (see Table 2). The generated ideas for a particular company were presented to the reviewer, organized by trend. The reviewer was then tasked with ranking each idea based on the metrics of *innovative, context* and *text*. Reviewers had the flexibility to adjust their scores during the evaluation of a single company. Once the evaluation for a company was completed, scores were finalized and could no longer be modified.

A reviewer was assigned to review all trends for a company. This approach enabled an accurate analysis of ideas in terms of their *context*, since all ideas for a company were considered collectively. This setup may raise concerns about the diversity of reviewers' perspectives for a single company, potentially impacting the validity of the experiment. We conducted an analysis of the average scores assigned by our reviewers, as depicted in Fig. 1. The graphic illustrates that none of the reviewers produced outlier scores. Although the ranges of scores for all categories are relatively wide, the metric of "innovation" exhibited the lowest range of quartiles. Based on this observation, we conclude that the assignment of reviewers to companies had minimal influence on the overall results.

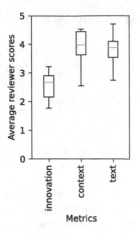

Fig. 1. Box plot of the average value that each reviewer assigned for the metrics.

Table 3. Average metric scores for all 70 ideas per company.

	Innovativeness	Context	Text
Average	**2.64**	**3.94**	**3.81**
Food	3.35	3.73	3.99
Automation	3.21	3.67	3.94
Hospitality	2.85	4.20	4.08
Non-profit	2.98	3.93	3.98
Spacecraft	3.19	3.95	4.02
Chemicals	2.49	4.20	3.86
IT	2.29	4.37	3.83
Law	2.34	4.14	3.88
Banking	1.77	3.58	3.47
Engineering	2.67	4.34	4.18
Legal	2.86	4.30	3.84
Education	2.82	3.57	3.62
Firearms	2.52	4.28	3.94
Laser	2.77	4.23	3.95
Nuclear	2.63	3.58	3.81
Recruitment	2.78	4.09	4.14
Entertainment	2.18	3.46	3.48
Healthcare	2.05	4.28	4.21
Social	2.43	3.63	3.49
Pharmaceuticals	2.44	3.74	3.32
Hospital	2.56	3.73	3.89
Media	2.44	4.14	3.93
Forestry	3.06	4.06	4.13
Sports	1.90	4.09	3.78
Dating	2.76	3.52	3.61
Software	2.44	3.81	3.54
Logistics	2.91	4.04	3.74
Insurance	2.61	3.55	3.50
Energy	2.77	4.14	3.73
Steel	2.90	4.03	3.82
Furniture	2.81	3.60	3.48

4   Results

We conducted an analysis of the ideas based on their average review scores per company, as presented in Table 3. The results show that the innovativeness is almost exactly the average, but the context and text quality generated via LLMs is rather high.

Table 4. Best and worst ranked *innovative* ideas.

Least innovative idea in **Firearms - Trend: Purpose-driven Marketing**: using the power of purpose to drive marketing decisions and strategies. Through this approach, companies can create a strong bond with their customers by understanding what motivates them and creating unique experiences that will meet those needs. The goal is to use purpose-driven marketing to build loyalty, trust and engagement with customers

---

Most innovative idea in **Firearms - Trend: Distributed Ledger**: creating a blockchain-based system that will enable users to securely store, track and transfer ownership of firearms. The Distributed Ledger technology created by *Firearms* is aimed at creating an immutable record of firearm transactions in order to improve safety and security for all involved parties. This new system will give law enforcement agencies the ability to trace weapons across their entire lifecycle, from production to purchase, sale, transport and storage. Additionally, it will provide real-time access to this data so that they can quickly identify suspicious activities or potential illegal activities related to firearms

---

Perception of Innovativeness: Was ranked the least high of all metrics. With an average score of 2.64 for all ideas, it suggests that the generated ideas fell in the middle range between not being innovative at all and being highly innovative. Certain industries or sectors, such as Banking and Sports, may inherently pose challenges in terms of generating innovative ideas. Banking is an established sector with limited scope for innovation, it is not surprising to observe lower scores. The sports industry is known for its continuous innovation, making it more difficult to identify ideas that have not already been explored.

Table 4 presents a selection of exemplary ideas, representing both the most promising and least innovative concepts overall. While several ideas received the same average score, we have chosen a single representative for each category. The least innovative idea related to *Firearms* (score of 0) is clearly lacking innovation, as it fails to offer any substantial new concept. In contrast, the notion of utilizing blockchain technology for weapons tracing stands out as a genuinely innovative proposal (score of 4.6). This idea not only highlights the potential applications of the technology, but also elucidates the resulting advantages it can yield.

The most innovative idea concerning *Firearms* was perceived as innovative by our reviewers. However, it is not entirely novel. Lokre et al. [12] describe a gun tracking system using blockchain technology in detail, and so do other publications [1,18]. Despite this, the idea itself was still perceived as novel.

A total of 7 ideas stood out as highly innovative, receiving an average score surpassing 4.5. Additionally, 97 ideas garnered a score of 4 or higher, making up approximately 4.47% of the corpus. Conversely, only 49 ideas were ranked with an innovativeness below 1, and a mere 3 ideas ranked below 0.5. Considering the average score of 2.64 across all ideas, it is evident that while Language Models (LLMs) like ours are not yet capable of true innovation, some ideas can be perceived as innovative by human readers who may not necessarily be domain experts. This was demonstrated in the case of the *Firearms* example.

Table 5. Best and worst fitting ideas in *context*.

---

*Least contextual idea in **Banking - Trend: Innovation transfer***: to create a new type of venture capital fund. This fund would focus on investing in innovative companies with the goal of transferring their technology and knowledge to other markets or sectors. The fund could be structured as an equity-based investment, debt financing, convertible notes, or any combination thereof. The fund would provide capital for early stage startups and established businesses that are working on cutting edge technologies that can be transferred to different markets and industries. Through this transfer of knowledge and technology, these businesses can gain access to larger customer bases and expand their reach into new areas of business opportunity

---

*Most context-fitting idea in **Energy - Trend: Energy planning as a school subject***: Introducing energy planning as a school subject. The goal of this project would be to educate the next generation on energy efficiency, renewable energy sources and their importance in our daily lives. This project would involve creating lesson plans for teachers around the topics of electricity, sustainability, hydropower, renewable energies and other related topics. *Energy* could provide resources such as technical information about their activities or experts from different areas to help create these lesson plans

---

Contextual Relevance: The *context* of the ideas received the highest average score of all 3 metrics, with a rating of 3.94. No company received a significantly negative score in this regard. This finding aligns with the known capabilities of LLMs, as they are designed to learn from and adapt to the provided context, enabling them to generate appropriate responses [20]. Research has shown that instructions on how to answer play a vital role in ensuring contextually faithful responses from LLMs [21,22]. Ideas ranked lower in terms of context, primarily focused on either the company for which the idea was intended or the specific trend the idea addressed, while neglecting the other aspect.

One of the ideas that ranked low in terms of context (see Table 5), pertains to the *banking* sector. Although the idea was generated within the context of *Innovation Transfer*, it primarily revolves around the concept of establishing a venture capital fund. While this idea is still relevant to the banking sector, it is more focused on investment funding rather than innovation transfer. In comparison, other ideas in the same sector proposed the creation of an innovation lab or a collaboration platform with universities, which align more closely with the intended context.

On the other hand, the most contextual idea, was specifically generated within the context of *Energy planning as a school subject*. This idea seamlessly integrates with the given context as it addresses the development of a lesson plan. Moreover, it aligns with the company's focus on providing sustainable energy solutions to its customers, further enhancing its contextual relevance.

Impressively, none of the ideas received a context ranking below 1. The poorest performing ideas obtained a ranking of 1.66. Out of the entire text corpus, a substantial 61.9% (1343 ideas) received a ranking of 4 or higher in terms of

Table 6. Best and worst *text* generated by the LLM.

Idea with bad text ranking in **Insurance - Trend: Cybersecurity**: a comprehensive Cybersecurity solution that provides cutting-edge technology and services to protect businesses from cyber threats. The solution consists of four main components: 1) Security Awareness Training: This component helps organizations understand the risks associated with cyber threats and how to mitigate them. It also provides employees with training on best practices for safe online behavior, such as secure password management, secure data storage, and application security

---

Idea with good text ranking in **Healthcare - Trend: Care home for seniors**: One concrete development that could raise sales numbers for *Healthcare* is the implementation of innovative technologies such as virtual reality (VR) and augmented reality (AR) to provide seniors with a more immersive care experience. VR and AR can be used to create interactive environments, allowing caregivers to provide better care while also providing seniors with an engaging way to stay connected with their families, friends, and activities outside of their homes. Additionally, these technologies can help reduce anxiety levels among elderly residents by creating calming visualizations or simulations

---

context. Furthermore, among these highly contextually relevant ideas, 354 ideas achieved a score of 4.5 or higher and were recognized as exceptionally relevant to the context for which the ideas were generated. This highlights the overall success in maintaining contextual coherence within the generated ideas.

Human-Like Text Quality: The evaluation of *text* quality yielded a slightly lower average score of 3.81 compared to the *context* ranking. No company received a low ranking in this aspect. Common shortcomings in the generated ideas included the lack of complete sentences at the beginning or end of the text. Worse-ranked ideas often exhibited more severe flaws, such as abruptly ending in the middle of a sentence or containing single-item lists, which compromised the overall cohesiveness of the text.

One of the worst ranked ideas with a value of 1.66 was identified in *Insurance* (see Table 6). The text starts with a lowercase character. The idea comprises four components, but only one of them is explained. While this limitation can be attributed to the token constraints during text generation, it clearly indicates that the text was generated rather than authored by a human.

One of several ideas ranked the maximum of 5 in *text*, is from *Healthcare*. The text exhibits grammatical correctness, introduces, and subsequently employs these abbreviations. The idea reads as a complete thought, without abruptly stopping in the middle of an explanation. This level of coherence and fluency further emphasizes the high-quality nature of the idea text.

Similarly to the *context* metric, it is notable that no ideas received a ranking below 1 in terms of *text* quality. 1343 ideas were ranked 4 or higher on average, indicating a generally good quality of generated text. 360 ideas (16.5%) received a score of 4.5 or higher, slightly more than in the context metric. This finding aligns with the advancements in language models, as it is becoming increasingly challenging to detect differences between human-generated and LLM-generated texts [19].

5   Conclusion and Outlook

Our analysis of 2170 texts generated by an LLM conclusively shows that an LLM is capable of most often generating *text* that is perceived as human equivalent, mostly fits a given *context*, and is capable of producing ideas that are only sometimes perceived *innovative*. These findings highlight the strengths and limitations of LLMs, showcasing their proficiency in mimicking human-like text while presenting the need for further advancements in generating innovative ideas.

Perception of Innovativeness: The study findings indicate that approximately 97 ideas, accounting for around 4.5% of the entire text corpus, were deemed highly innovative by human reviewers. Conversely, only 49 ideas were ranked as generic or lacking innovativeness. Considering the context of innovation management and the reduced effort required for an innovation manager to review and evaluate paragraph-long ideas, this level of innovativeness demonstrated by the LLM justifies its use case.

Contextual Relevance: A majority of 1343 ideas, or 61.9% of the total, were ranked as fitting the context, encompassing both the company and the trend for which the ideas were generated. Interestingly, in some instances, the LLM overlooked either the context of the given company or the context of the trend, resulting in lower scores. No idea was completely mismatched with the provided context.

Human-Like Text Quality: The evaluation revealed that 61.9% of the entire text corpus received rankings indicative of human-like text quality, with no text being deemed completely faulty. This outcome aligns with expectations, as LLMs have demonstrated their ability to generate text that closely resembles human-written content.

Our study on the innovativeness of LLM-generated texts within a given context has provided conclusive insights. However, it is important to note that there are still many aspects to consider when evaluating LLMs. Existing benchmark frameworks like the Holistic Evaluation of Language Model (HELM) Benchmark [11] encompass various metrics relevant metrics such as fairness and toxicity, but they may not cover task-specific metrics such as *innovativeness*, which is highly relevant in innovation management.

In the field of innovation management, there is still much work to be done. One potential avenue is exploring unsupervised methods for guiding the idea space using LLMs to conduct idea ranking, similar to the process undertaken by our human reviewers. This approach could enable the automatic ranking or filtering of ideas based on their scores, thereby enhancing the presentation of novel ideas to human innovation managers.

Acknowledgements and Funding. Funding was provided by the Austrian Research Promotion Agency (FFG) under the Project Explainable Creativity (EACI, project number 892004).

We thank AnyIdea (https://anyidea.ai/) for their provision of data sets used in this work. We thank the project partner Cloudflight (https://www.cloudflight.io/), especially Michael Weissenböck, Anna Hausberger and Rine Rajendran, for their valuable contributions to this work. We thank Michaela Jungwirth and Marius Elflein for the conceptualization and the organization of the evaluation, and the reviewers of the texts for their valuable contribution to this work.

References

1. Akello, P., Vemprala, N., Lang Beebe, N., Raymond Choo, K.K.: Blockchain use case in ballistics and crime gun tracing and intelligence: toward overcoming gun violence. ACM Trans. Manage. Inf. Syst. **14**(1), 1–26 (2022). https://doi.org/10.1145/3571290

2. Bender, E.M., Gebru, T., McMillan-Major, A., Shmitchell, S.: On the dangers of stochastic parrots: can language models be too big? In: Proceedings of the 2021 ACM Conference on Fairness, Accountability, and Transparency, pp. 610–623. ACM (2021). https://doi.org/10.1145/3442188.3445922

3. Bouschery, S.G., Blazevic, V., Piller, F.T.: Augmenting human innovation teams with artificial intelligence: Exploring Transformer-Based Language Models. J. Product Innov. Manage. **40**(2), 139–153 (2023). https://doi.org/10.1111/jpim.12656

4. Brown, T.B., et al.: Language models are few-shot learners. CoRR arXiv:2005.14165 (2020)

5. Chowdhery, A., et al.: PaLM: scaling language modeling with pathways. arXiv:2204.02311 (2022)

6. Cooper, R.G., Edgett, S.: Ideation for product innovation. What are the Best Methods, pp. 12–17 (2008)

7. Del Vecchio, P., Di Minin, A., Petruzzelli, A.M., Panniello, U., Pirri, S.: Big data for open innovation in SMEs and large corporations: trends, opportunities, and challenges. Creat. Innov. Manage. **27**(1), 6–22 (2018). https://doi.org/10.1111/caim.12224

8. Di Fede, G., Rocchesso, D., Dow, S.P., Andolina, S.: The idea machine: LLM-based expansion, rewriting, combination, and suggestion of ideas. In: Creativity and Cognition, pp. 623–627. C&C 2022, Association for Computing Machinery. https://doi.org/10.1145/3527927.3535197

9. Greshake, K., Abdelnabi, S., Mishra, S., Endres, C., Holz, T., Fritz, M.: More than you've asked for: a comprehensive analysis of novel prompt injection threats to application-integrated large language models (2023). arXiv:2302.12173

10. Krauss, O., Aschauer, A., Stöckl, A.: Modelling shifting trends over time via topic analysis of text documents. https://doi.org/10.46354/i3m.2022.mas.009. https://www.cal-tek.eu/proceedings/i3m/2022/mas/009

11. Liang, P., et al.: Holistic evaluation of language models (2022)

12. Lokre, S.S., Naman, V., Priya, S., Panda, S.K.: Gun tracking system using blockchain technology. In: Panda, S.K., Jena, A.K., Swain, S.K., Satapathy, S.C. (eds.) Blockchain Technology: Applications and Challenges. ISRL, vol. 203, pp. 285–300. Springer, Cham (2021). https://doi.org/10.1007/978-3-030-69395-4_16
13. McCoy, R.T., Smolensky, P., Linzen, T., Gao, J., Celikyilmaz, A.: How much do language models copy from their training data? Evaluating linguistic novelty in text generation using RAVEN (2021). arXiv:2111.09509
14. Min, B., et al.: Recent advances in natural language processing via large pre-trained language models: a survey. CoRR arXiv:2111.01243 (2021)
15. Radford, A., et al.: Language models are unsupervised multitask learners. OpenAI Blog **1**(8), 9 (2019)
16. Sarsa, S., Denny, P., Hellas, A., Leinonen, J.: Automatic generation of programming exercises and code explanations using large language models. In: Proceedings of the 2022 ACM Conference on International Computing Education Research - Volume 1. ICER 2022, vol. 1, pp. 27–43. Association for Computing Machinery. https://doi.org/10.1145/3501385.3543957
17. Scao, T.L., et al.: Bloom: A 176b-parameter open-access multilingual language model. arXiv preprint arXiv:2211.05100 (2022)
18. Soni, D.B., Mahler, M.L.: Blockchain technology for a firearm registry. New Zealand J. Business Technol. **2**, 34–42 (2020)
19. Tang, R., Chuang, Y.N., Hu, X.: The science of detecting LLM-generated texts. arXiv:2303.07205
20. Wang, X., Zhu, W., Wang, W.Y.: Large language models are implicitly topic models: explaining and finding good demonstrations for in-context learning (2023)
21. White, J., et al.: A prompt pattern catalog to enhance prompt engineering with ChatGPT. arXiv:2302.11382
22. Zhou, W., Zhang, S., Poon, H., Chen, M.: Context-faithful Prompting for Large Language Models. arXiv:2303.11315

LocBERT: Improving Social Media User Location Prediction Using Fine-Tuned BERT

Asif Khan[1] ⓘ, Huaping Zhang[1]([✉]) ⓘ, Nada Boudjellal[2]([✉]) ⓘ, Arshad Ahmad[3] ⓘ, and Maqbool Khan[3] ⓘ

[1] School of Computer Science and Technology, Beijing Institute of Technology, Beijing 100081, China
kevinzhang@bit.edu.cn
[2] The Faculty of Information and Communication Technology, University Abdel-Hamid Mehri Constantine 2, 25000 Constantine, Algeria
Nada.boudjellal@univ-constantine2.dz
[3] Department of IT and Computer Science, Pak-Austria Fachhochschule: Institute of Applied Sciences and Technology, Mang Khanpur Road, Haripur 22620, Pakistan

Abstract. Predicting user locations on social media platforms like Twitter is a challenging task with numerous applications in marketing, politics, and disaster response. This paper introduces LocBERT, a fine-tuned BERT model designed to accurately predict the locations of Twitter users based on their conversations. Our experiments focus on the "StateElecTweets" dataset, comprising 1.6 million labeled tweets associated with state locations within the United States. The results demonstrate that LocBERT outperforms traditional machine learning models such as SVM and Naive Bayesian, achieving an accuracy of 0.988 and an F1-score of 0.987. The study contributes to predicting the location of users in election-related tweets, enabling a better understanding of campaign demographics and assisting political stakeholders in refining their strategies. The findings of this research hold significant implications for various domains and highlight the effectiveness of LocBERT in accurately predicting user locations on social media platforms.

Keywords: Social Media · Transformers · Location · Machine Learning

1 Introduction

Social media usage has exploded over the past decade, with 4.76 billion people using social media platforms worldwide in January 2023 [1]. This represents a staggering 59.4% of the global population. The advent of social media platforms, particularly Twitter, has revolutionized the way we access information and communicate with each other. During elections, these platforms play a pivotal role as primary sources for gathering insights and understanding public sentiment [2–4]. The ability to study and analyze election predictions on social media has emerged as a critical aspect for researchers, politicians, and decision-makers in their quest to comprehend the pulse of the people. Analyzing election campaigns at the state and regional level has become increasingly crucial, as it enables a deeper understanding of the electorate's preferences and sentiments [5].

G. Kotsis et al. (Eds.): DEXA 2023 Workshops, CCIS 1872, pp. 23–32, 2023.
https://doi.org/10.1007/978-3-031-39689-2_3

Predicting election outcomes at the state or regional level using social media data can provide valuable insights into the dynamics of election campaigns [6]. However, accurately identifying the location of Twitter users during campaigns poses a significant challenge due to the lack of location information provided by a considerable number of users. To address this challenge and gain a better understanding of the political landscape, we draw inspiration from successful applications of BERT models [12] in various domains. These include advancements in patent classification [7], scientific text analysis with SciBERT [8], financial text processing using FinBERT [9], French language tasks with CamemBERT [10], and disease and treatment-named entity recognition in Arabic biomedical texts with ABioNER [11]. Additionally, we leverage the versatility of BERT models and introduce LocBERT, which is fine-tuned on a vast dataset of 1.6 million election-related tweets, each labelled with the corresponding state's location in the United States. By harnessing the complexity of language models, LocBERT enhances prediction capabilities and enables accurate identification of the location associated with Twitter users in election-related tweets.

This study encompasses two primary contributions. First, we present the "state-ElecTweets dataset," comprising 1.6 million tweets, where each tweet is associated with the user's location at the state level within the United States. This dataset serves as a valuable resource for researchers and practitioners interested in studying election-related discourse on Twitter. Second, we introduce LocBERT, a fine-tuned BERT model that specializes in predicting the locations of Twitter users within election-related tweets.

The implications of this study extend beyond academic research. Policymakers, politicians, and decision-makers can leverage the insights obtained from LocBERT to gain a deeper understanding of public sentiment and tailor their strategies accordingly. By harnessing the power of social media data, this work has the potential to influence political campaigns and policy decisions, ultimately leading to more informed and effective political engagement.

The subsequent sections of this paper are structured as follows: Sect. 2 presents a comprehensive review of the relevant literature in the field. In Sect. 3, we provide an overview of our dataset and describe the methodology employed for training and fine-tuning LocBERT. The experimental setup and results are presented in Sect. 4, where we discuss and analyze the outcomes. Finally, we conclude this study in the concluding section, summarizing the key findings and their implications.

2 Related Work

Several studies have addressed the problem of location prediction for Twitter users and tweets. Chong & Lim, [13] proposed a learning-to-rank framework that incorporated contextual information such as tweet posting time and user location history, improving the accuracy of venue ranking. In a similar vein, Flatow et al. [14] utilized n-gram location distributions to identify the location of non-geotagged social media content, considering the trade-off between accuracy and coverage across platforms and devices.

To estimate tweet locations, Priedhorsky et al. [15] introduced a scalable, content-based approach based on Gaussian mixture models. Their approach demonstrated reliable and well-calibrated estimates, emphasizing the importance of toponyms and languages with a small geographic footprint. Additionally, Miura et al. [16] proposed a geolocation prediction model using a complex neural network, outperforming previous ensemble approaches and capturing the statistical characteristics of the datasets.

In addition to predicting tweet locations, researchers have explored predicting Twitter users' home locations. Chang et al. [17] compared probability models and proposed unsupervised methods to remove noisy data, achieving comparable results to state-of-the-art methods. Similarly, Mahmud et al. [18] presented an algorithm that incorporated statistical and heuristic classifiers, achieving improved accuracy by analyzing user movement patterns and leveraging a geographic gazetteer dictionary.

Furthermore, several studies have investigated city-level geolocation prediction systems. Han et al. [19] implemented a stacking approach combining tweet text and user-declared metadata, achieving higher accuracy compared to benchmark methods. Their study highlighted the impact of temporal factors on model generalization and discussed potential applications of the system. Al Hasan Haldar et al. [20] addressed location prediction from implicit information in social networks and evaluated eight prediction models across real-world datasets, providing valuable insights into the strengths and limitations of these models, contributing to a better understanding of the location prediction problem.

Specifically in the context of Indonesian tweets, Simanjuntak et al. [21] utilized machine learning approaches such as LSTM and BERT to predict tweet locations. By combining user information and aggregated tweet content, their proposed model achieved superior accuracy compared to baseline models.

Though these studies have made significant contributions to the field of location prediction for Twitter users and tweets. However, there is still a need for region/state-level location prediction. In our study, we present LocBERT, a novel approach that fine-tunes BERT on a dataset of 1.61 million tweets with user location information provided as states of the USA. By incorporating user-related features and tweet content, the model utilizes the power of BERT to accurately predict the state location of Twitter users. This targeted approach allows for valuable insights into user demographics, regional preferences, and engagement related to elections. LocBERT contributes to the field by addressing the specific challenge of state-level prediction in the context of Twitter data.

3 Predicting User Location

In this section, we will present an overview of our dataset, "StateElectTweets," and outline the methodology used for our LocBERT model.

3.1 Data

We collected tweets using the Twitter Search API for our dataset. Our dataset comprises tweets related to the Democratic and Republican parties, identified through hashtags such as #Republican, #GOP, #Democratic, and #TheDemocrats. These tweets were collected from December 2019 to November 2020. Additionally, we gathered tweets mentioning Joe Biden and Donald Trump using hashtags like #realdonaldtrump, #Trump2020, #JoeBiden, and #Biden2020. This subset of tweets was collected from July 2020 to November 2020.

Twitter offers users the option to provide location information while tweeting, either as precise geo-location or general location details in their user profile. Precise geo-location provides specific latitude and longitude coordinates, while general location includes information like country, state, or city. Accessing this information can be done through the Twitter API, using fields such as "geo" or "coordinates." However, since June 18, 2019, Twitter has limited access to precise coordinates and instead provides more generalized location data such as country, region, and city.

For our dataset, we only used tweets that provided user location information. We discarded tweets that did not include location information. Our "Supp-Loc dataset" contains 3,432,379 tweets with user location information. Among these tweets, 317,454 users provided their location as the United States, including mentions of USA, America, or US. Furthermore, 1,610,694 tweets mentioned specific state names within the USA, such as NY, New York, Florida, California, or CA. The remaining 1,503,601 tweets mentioned locations outside of the USA, such as Canada, Spain, and others.

```
1 states = ['Alabama', 'Alaska', 'Arizona', 'Arkansas', 'California', 'Colorado', 'Connecticut', 'Delaware', 'Florida', 'Georgia', 'Hawaii',
2     'Idaho', 'Illinois', 'Indiana', 'Iowa', 'Kansas', 'Kentucky', 'Louisiana', 'Maine', 'Maryland', 'Massachusetts', 'Michigan', 'Minnesota',
3     'Mississippi', 'Missouri', 'Montana', 'Nebraska', 'Nevada', 'New Hampshire', 'New Jersey', 'New York', 'New Mexico', 'North Carolina',
4     'North Dakota', 'Ohio', 'Oklahoma', 'Oregon', 'Pennsylvania', 'Rhode Island', 'South Carolina', 'South Dakota', 'Tennessee', 'Texas', 'Utah',
5     'Vermont', 'Virginia', 'Washington', 'West Virginia', 'Wisconsin', 'Wyoming']
6 stateCodes = ['AL', 'AK', 'AZ', 'AR', 'CA', 'CO', 'CT', 'DE', 'FL', 'GA', 'HI', 'ID', 'IL', 'IN', 'IA', 'KS', 'KY', 'LA', 'ME', 'MD', 'MA', 'MI', 'MN',
7     'MS', 'MO', 'MT', 'NE', 'NV', 'NH', 'NJ', 'NM', 'NY', 'NC', 'ND', 'OH', 'OK', 'OR', 'PA', 'RI', 'SC', 'SD', 'TN', 'TX', 'UT', 'VT', 'VA',
8     'WA', 'WV', 'WI', 'WY']
9 stateMapping = {'AL': 'Alabama', 'AK': 'Alaska', 'AZ': 'Arizona', 'AR': 'Arkansas', 'CA': 'California',
10     'CO': 'Colorado', 'CT': 'Connecticut', 'DE': 'Delaware', 'FL': 'Florida', 'GA': 'Georgia',
11     'HI': 'Hawaii', 'ID': 'Idaho', 'IL': 'Illinois', 'IN': 'Indiana', 'IA': 'Iowa', 'KS': 'Kansas',
12     'KY': 'Kentucky', 'LA': 'Louisiana', 'ME': 'Maine', 'MD': 'Maryland', 'MA': 'Massachusetts',
13     'MI': 'Michigan', 'MN': 'Minnesota', 'MS': 'Mississippi', 'MO': 'Missouri', 'MT': 'Montana',
14     'NE': 'Nebraska', 'NV': 'Nevada', 'NH': 'New Hampshire', 'NJ': 'New Jersey', 'NY': 'New York',
15     'NM': 'New Mexico', 'NC': 'North Carolina', 'ND': 'North Dakota', 'OH': 'Ohio', 'OK': 'Oklahoma',
16     'OR': 'Oregon', 'PA': 'Pennsylvania', 'RI': 'Rhode Island', 'SC': 'South Carolina',
17     'SD': 'South Dakota', 'TN': 'Tennessee', 'TX': 'Texas', 'UT': 'Utah', 'VT': 'Vermont',
18     'VA': 'Virginia', 'WA': 'Washington', 'WV': 'West Virginia', 'WI': 'Wisconsin', 'WY': 'Wyoming'}
19 stateString = ['Alabama', 'Alaska', 'Arizona', 'Arkansas', 'California', 'Colorado', 'Connecticut', 'Delaware', 'Florida', 'Georgia', 'Hawaii',
20     'Idaho', 'Illinois', 'Indiana', 'Iowa', 'Kansas', 'Kentucky', 'Louisiana', 'Maine', 'Maryland', 'Massachusetts', 'Michigan', 'Minnesota',
21     'Mississippi', 'Missouri', 'Montana', 'Nebraska', 'Nevada', 'New Hampshire', 'New Jersey', 'New York', 'New Mexico', 'North Carolina',
22     'North Dakota', 'Ohio', 'Oklahoma', 'Oregon', 'Pennsylvania', 'Rhode Island', 'South Carolina', 'South Dakota', 'Tennessee', 'Texas', 'Utah',
23     'Vermont', 'Virginia', 'Washington', 'West Virginia', 'Wisconsin', 'Wyoming', 'AL', 'AK', 'AZ', 'AR', 'CA', 'CO', 'CT', 'DE', 'FL', 'GA', 'HI', 'ID', 'IL'
24     'MS', 'MO', 'MT', 'NE', 'NV', 'NH', 'NJ', 'NM', 'NY', 'NC', 'ND', 'OH', 'OK', 'OR', 'PA', 'RI', 'SC', 'SD', 'TN', 'TX', 'UT', 'VT', 'VA',
25     'WA', 'WV', 'WI', 'WY']
```

Fig. 1. States (USA) dictionary

In this study, we utilized the "StateElecTweets" dataset, which is a subset of the larger "Supp-Loc dataset. The StateElecTweets dataset comprises 1,610,694 tweets that specifically mention locations within the United States. To associate each tweet with a particular U.S. state, we employed a dictionary that matched the state names with the users' mentioned locations. For instance, if a user stated their location as "NY, USA" or "New York," both would be categorized under the state of New York after applying the states-dictionary. The mapping of states to their corresponding names is illustrated in Fig. 1.

3.2 LocBERT

In this study, we proposed "LocBERT" – location prediction BERT-based model. We used our *StateElecTweets,* which contains tweets from the states only. To train our

LocBERT model, we include user ID, user screen name, user name, user location, state, and text as features in our training data. The number of states in our dataset is imbalanced. To address the imbalanced dataset in this study, we utilized the SMOTE (Synthetic Minority Over-sampling Technique) method. SMOTE helps balance the number of instances for each state in our dataset by generating synthetic samples for the minority class states. By doing so, we ensure that our LocBERT model receives adequate training data for every state, allowing for accurate predictions across all states. Before training our LocBERT, we needed to pre-process the data to prepare it for use with a BERT-based model. We pre-processed the data to ensure that it was in a format suitable for our task. First, we performed several text pre-processing steps. We then tokenized the text of each tweet using the BERT tokenizer, which splits the text into a sequence of word pieces. We took a BERT model and fine-tuned it using the Hugging Face Transformers library. Figure 2 shows the fine-tuning process of our LocBERT model. The following is our approach for LocBERT.

Let X_1, X_2,..., X_n be the input batch of n tweets,

Where $X = $ *(text, user name, user screen name, user id, user location, state)* be the input to the model.

Where the state is the state name of USA assigned after applying the state-dictionary.

Let Y_1, Y_2,..., Y_n be their corresponding labels, where each Y_i is a set of unique integer values representing the "state" for the ith example in the batch.

We pass batch X through the pre-trained BERT model to obtain the output H, which is a sequence of hidden states for each example in the batch. We apply a pooling operation on H to obtain a fixed-sized representation C of the input X for each example in the batch.

We add a classification layer on top of C, which is a fully-connected layer with n neurons (where n is the number of unique location labels, which is 51). Let W be the weight matrix of the classification layer, and b be the bias vector.

We compute the output O for each example in the batch as:

$$O_i = exp(W_i C_i + b_i) \Sigma_i exp\left(W_j C_i + b_j\right) \tag{1}$$

where i is the index of the example in the batch, W_i is the ith row of the weight matrix W, b_j is the ith element of the bias vector b, and $\sum_j exp(W_j c_i + b_j)$ is the sum of exponential scores for all possible labels j. The softmax function ensures that the output probabilities sum to 1, and that the highest probability corresponds to the most likely label. In other words, the softmax function assigns a probability to each state, indicating the likelihood that the user belongs to that state.

We define the loss function L for the batch as the mean cross-entropy loss between O and Y:

$$L = -(1/n) \Sigma_i \Sigma_{ji} y_{ij} log\left(O_{ij}\right) \tag{2}$$

where y_{ij} is the jth element of the one-hot encoded vector of Y_i for the ith example in the batch, and O_{ij} is the jth element of the output vector O_i for the ith example in the batch.

We optimize the model by minimizing L using the AdamW optimizer with a learning rate of 5e-5. This optimizer updates the model's weights based on the gradient of the loss function, which measures the difference between the predicted and actual labels.

During training, we use mini-batch gradient descent with a batch size of 32, and train the model for a total of 10 epochs.

Finally, we evaluate the performance of our model on the test set of tweets and location labels. We use several metrics to assess the accuracy of our predictions, including accuracy, precision, recall, and F1-score. We also perform a qualitative analysis of the model's predictions, examining a sample of tweets and their corresponding predicted locations to gain insight into the model's strengths and weaknesses.

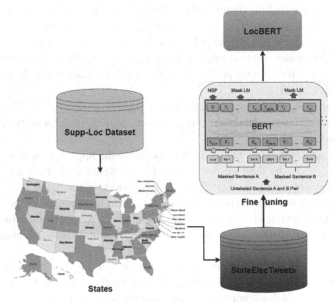

Fig. 2. Overview of LocBERT

4 Experimental Results and Discussion

The experiments were conducted on an NVIDIA Titan XP GPU with 12GB VRAM and a system equipped with 128GB of RAM. The performance evaluation of LocBERT was carried out using different metrics including accuracy, precision, recall, and F1-score.

We trained LocBERT, SVM, and Naïve Bayesian (NB) models on our "State-ElecTweets" dataset to predict the state of Twitter users. LocBERT is a language model that was fine-tuned specifically for this task. The LocBERT model was trained using two different approaches:

1. LocBERT_without_location
2. LocBERT_with_location

The LocBERT model was fine-tuned for 10 epochs, utilizing the AdamW optimizer with a batch size of 32 and a learning rate of 5e-5. The dataset was split into a training set (80%) and a testing set (20%) to train and evaluate the models, respectively. The first model, "LocBERT_without_location", consists of features; user ID, user name, user screen name, text, and state. We exclude "user location" feature from our dataset to observe the performance of LocBERT. As some tweets still contain information that can assist the language model in predicting the location. For example, a tweet from a user named "NYC Sights Sounds" would indicate a location in New York. Figure 3 shows the evaluation metrics for the "LocBERT_without_location" model. Although the overall performance of the model is not so well the figure indicates that the performance of the model improved gradually over the epochs. The accuracy increased from 0.166 at the beginning of training to 0.3158 after the 10th epoch, indicating that the model is able to correctly predict the states with increasing accuracy as it is trained on more. Similarly, precision and recall both showed improvement over the training steps. Precision increased from 0.26 to 0.51, indicating that the model is making fewer false positive predictions as it is fine-tuned. Recall increased from 0.07 to 0.23, indicating that the model is identifying more true positives as it is trained on more data. The F1 score, which is a harmonic mean of precision and recall, also increased from 0.09 to 0.30, indicating that the model's ability to balance between precision and recall improved over the epochs.

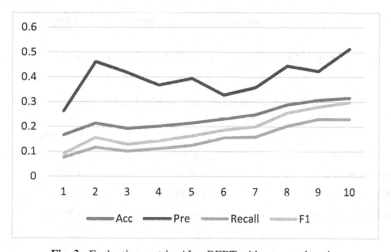

Fig. 3. Evaluation metrics I LocBERT without user_location

In contrast, the second version of the LocBERT model, called "LocBERT_with_location", included the "user location" feature in addition to the other features such as user ID, user name, user screen name, text, and state. The inclusion of this feature is for the evaluation of the model's performance and analysis of the impact of the "user location" feature. Figure 4 depicts the evaluation metrics for the "LocBERT_with_location" model. Interestingly, this model achieved much higher performance scores as compared to "LocBERT_without_location", with accuracy scores ranging from 0.9503 to 0.9879. We chose the model with 0.9879 accuracy (See Table 1), considering that the model correctly predicted the class of approximately 98.79% of the tweets in the dataset. During training, the precision score for it ranged from 0.9516 to 0.988004, and we chose the model's best performance, where the precision was 0.988804, indicating that out of all the tweets predicted as belonging to a certain class, 98.80% were actually in that class. The recall score for this model was 0.987693 at its best, which indicates that out of all the tweets in a certain class, 98.77% were correctly identified by the model. Finally, the F1-score for this model was 0.987846.

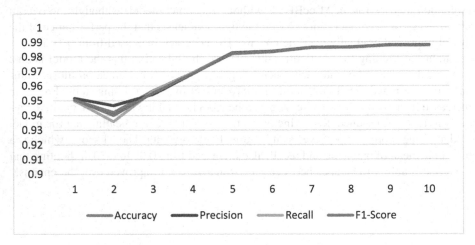

Fig. 4. Evaluation metrics l LocBERT with user_location

Table 1. Evaluation Metrics for LocBERT, SVM, and NB

Models	Accuracy	F1-Score
SVM-with user_loc	0.421	0.35
Naïve Bayesian-with user_loc	0.395	0.379
LocBERT-without user_loc	0.3158	0.298
LocBERT-with user_loc	0.988	0.987

Furthermore, we trained Support Vector Machines (SVM) and Naïve Bayesian (NB) models and compared our model's results with them. We trained these models by including "user location". Table 1 shows the accuracy and F1 score of LocBERT along with SVM and NB. For SVM, we observed an accuracy of 0.421 and an F1-score of 0.35. Similarly, NB, trained with user location, achieved an accuracy of 0.395 and an F1-score of 0.379. These results show that the "LocBERT_with_location" model outperformed other models including "LocBERT_without_location" by a wide margin. This suggests that including user location information in training improves performance. It is hard to predict the state just from the user ID, user name, and text.

5 Conclusion

In this study, we presented LocBERT, a fine-tuned BERT model designed to predict the locations of Twitter users based on their conversations. Our experiments on the state-ElecTweets dataset, comprising 1.6 million tweets labeled with corresponding state locations within the United States, demonstrated the effectiveness of LocBERT in accurately identifying user locations. The evaluation metrics highlighted the strong performance of LocBERT, particularly when incorporating the "user location" feature in the training process. Our results outperformed other models, including SVM and Naïve Bayesian, emphasizing the importance of user location information for accurate prediction.

However, despite the promising results, our study has certain limitations. Firstly, our focus was limited to the United States, and the effectiveness of LocBERT in predicting user locations in other countries remains unexplored. Secondly, this study focused on comparing LocBERT with traditional machine learning models, such as SVM and Naïve Bayesian, and did not include a comparison with other deep learning models like LSTM. Including a comprehensive analysis of various deep learning architectures could provide further insights into the performance and suitability of different models for location prediction.

Future work should aim to expand the geographical scope of LocBERT by incorporating datasets from different countries, enabling the prediction of user locations on a global scale. Moreover, enhancing LocBERT's capabilities to handle incomplete or missing location information would be a valuable direction for further research. This could involve leveraging external geospatial data or exploring advanced natural language processing techniques to extract contextual cues that indicate user locations.

Furthermore, extending LocBERT beyond election-related tweets to other domains of interest, such as public opinion analysis or sentiment classification, could provide valuable insights into various socio-political contexts. By broadening its application, LocBERT has the potential to contribute to a deeper understanding of user behaviour and sentiment across different topics and domains.

References

1. "Global Social Media Statistics — DataReportal – Global Digital Insights." https://datarepor tal.com/social-media-users. Accessed 26 Feb 2023
2. Santos, J.S., Bernardini, F., Paes, A.: A survey on the use of data and opinion mining in social media to political electoral outcomes prediction. Soc. Netw. Anal. Min. 11(1), 1–39 (2021). https://doi.org/10.1007/s13278-021-00813-4
3. Khan, A., et al.: Predicting politician's supporters' network on twitter using social network analysis and semantic analysis. Sci. Program. 2020, 1–17 (2020). https://doi.org/10.1155/2020/9353120
4. Chauhan, P., Sharma, N., Sikka, G.: The emergence of social media data and sentiment analysis in election prediction. J. Ambient. Intell. Humaniz. Comput. 12(2), 2601–2627 (2020). https://doi.org/10.1007/s12652-020-02423-y
5. Khan, A., et al.: Election prediction on twitter: a systematic mapping study. Complexity 2021, 1–27 (2021). https://doi.org/10.1155/2021/5565434

6. Heredia, B., Prusa, J.D., Khoshgoftaar, T.M.: Location-based twitter sentiment analysis for predicting the U.S. 2016 presidential election. In: Proceedings of the 31st International Florida Artificial Intelligence Research Society Conference, FLAIRS 2018, vol. 2009, pp. 265–270 (2018)
7. Lee, J.S., Hsiang, J.: Patent classification by fine-tuning BERT language model (2020)
8. Beltagy, I., Lo, K., Cohan, A.: SCIBERT: a pretrained language model for scientific text. In: EMNLP-IJCNLP 2019 - 2019 Conference on Empirical Methods in Natural Language Processing and 9th International Joint Conference on Natural Language Processing, Proceedings of the Conference, pp. 3615–3620 (2019). https://doi.org/10.18653/v1/d19-1371
9. Huang, A.H., Wang, H., Yang, Y.: FinBERT: a large language model for extracting information from financial text*. Contemp. Account. Res. (2022). https://doi.org/10.1111/1911-3846.12832
10. Martin, L., et al.: CamemBERT: a Tasty french language model, pp. 7203–7219 (2020). https://doi.org/10.18653/v1/2020.acl-main.645
11. Boudjellal, N., et al.: ABioNER: a BERT-based model for arabic biomedical named-entity recognition. Complexity **2021**, 1–6 (2021). https://doi.org/10.1155/2021/6633213
12. Devlin, J., Chang, M.W., Lee, K., Toutanova, K.: BERT: pre-training of deep bidirectional transformers for language understanding. In: NAACL HLT 2019 - 2019 Conference North North American Chapter of the Association for Computational Linguistics: Human Language Technologies - Proceedings Conference, vol. 1, pp. 4171–4186 (2019)
13. Chong, W.H., Lim, E.P.: Exploiting contextual information for fine-grained tweet geolocation. In: Proceedings of the 11th International Conference on Web and Social Media, ICWSM 2017, pp. 488–491 (2017). https://doi.org/10.1609/icwsm.v11i1.14909
14. Flatow, D., Naaman, M., Xie, K.E., Volkovich, Y., Kanza, Y.: On the accuracy of hyper-local geotagging of social media content. In: WSDM 2015 - Proceedings of the 8th ACM International Conference on Web Search and Data Mining, pp. 127–136 (2015). https://doi.org/10.1145/2684822.2685296
15. Priedhorsky, R., Culotta, A., Del Valle, S.Y.: Inferring the origin locations of tweets with quantitative confidence. In: Proceedings of the ACM Conference on Computer Supported Cooperative Work, CSCW, pp. 1523–1536 (2014). https://doi.org/10.1145/2531602.2531607
16. Miura, Y., Taniguchi, T., Taniguchi, M., Ohkuma, T.: Unifying text, metadata, and user network representations with a neural network for geolocation prediction. In: ACL 2017 - 55th Annual Meeting of the Association for Computational Linguistics, Proceedings of the Conference (Long Papers), vol. 1, pp. 1260–1272 (2017).https://doi.org/10.18653/v1/P17-1116
17. Chang, H.W., Lee, D., Eltaher, M., Lee, J.: Phillies tweeting from philly? Predicting twitter user locations with spatial word usage. In: Proceedings of the 2012 IEEE/ACM International Conference on Advances in Social Networks Analysis and Mining, ASONAM 2012, pp. 111–118 (2012). https://doi.org/10.1109/ASONAM.2012.29
18. Mahmud, J., Nichols, J., Drews, C.: Home location identification of twitter users. ACM Trans. Intell. Syst. Technol. **5**(3) (2014). https://doi.org/10.1145/2528548
19. Han, B., Cook, P., Baldwin, T.: A stacking-based approach to Twitter user geolocation prediction. In: Proceedings of the Annual Meeting of the Association for Computational Linguistics, vol. 2013-Augus, pp. 7–12 (2013)
20. Al Hasan Haldar, N., Li, J., Reynolds, M., Sellis, T., Yu, J.X.: Location prediction in large-scale social networks: an in-depth benchmarking study. VLDB J. **28**(5), 623–648 (2019). https://doi.org/10.1007/s00778-019-00553-0
21. Simanjuntak, L.F., Mahendra, R., Yulianti, E.: We know you are living in bali: location prediction of twitter users using BERT language model. Big Data Cogn. Comput. **6**(3) (2022). https://doi.org/10.3390/bdcc6030077

Measuring Overhead Costs of Federated Learning Systems by Eavesdropping

Rainer Meindl[1]($\boxtimes$) and Bernhard A. Moser[1,2]

[1] Software Competence Center Hagenberg (SCCH), Softwarepark 32a,
4232 Hagenberg, Austria
{rainer.meindl,bernhard.moser}@scch.at
[2] Institute of Signal Processing, Johannes Kepler University of Linz, Linz, Austria
bernhard.moser@jku.at

Abstract. This paper addresses the issue of communication overhead costs of federated learning including transmission bandwidth and synchronisation efforts. These costs typically consist of locally observable costs on executing components, but there are also hidden costs that can only be measured from a system-wide perspective. The goal is to provide insight into these hidden costs, measure them and identify strategies for reducing them. We propose an approach to tackle the hidden costs by establishing a methodology consisting of an eavesdropping concept and an evaluation strategy. This way we obtain a refined analysis of directly observable costs contrasting hidden costs, which is underpinned by experiments based on a 40-client-spanning federated learning system and the FEMNIST dataset.

Keywords: Federated Learning · Optimisation · Privacy-Preserving · Transparency

1 Introduction

Federated learning is a distributed machine learning approach that enables the training of a model on decentralized data. This approach has gained popularity recently due to its ability to handle sensitive data while maintaining privacy and security [3,12]. However, the overhead cost of federated learning is an important consideration, as it involves the transmission of large amounts of data over a network, as well as the time spent synchronizing the clients [6]. In this paper, we explore the cost of federated learning, including the directly observable cost in terms of bytes generated, as well as hidden costs, such as bytes transmitted over a network. Our goal is to provide a comprehensive understanding of the overhead cost of federated learning and to identify potential strategies for reducing this cost. This includes the definition of this overhead cost, how and where we can measure it, the influence of the measurement towards the system and finally, the scalability of this approach to large-scale federated learning systems.

G. Kotsis et al. (Eds.): DEXA 2023 Workshops, CCIS 1872, pp. 33–42, 2023.
https://doi.org/10.1007/978-3-031-39689-2_4

The paper is structured as follows. First, we provide a comprehensive overview of the related work in Sect. 2, highlighting key contributions and limitations of previous research. In Sect. 3 we present the methodology and techniques used in our study. We also discuss the rationale behind our design choices and how they contribute to achieving our research goal. Section 4 provides details about a reference implementation, including used libraries, hardware specifications, software infrastructure, data sets, etc. We discuss the challenges we encountered during our implementation and application of the novel framework in Sect. 5. It is followed by the results in Sect. 6. We conclude with Sect. 7 in which we highlight the potential areas for future research, including possible extensions of the framework.

2   Related Work

Many studies have focused on optimizing the performance of federated learning algorithms to minimize computational and communication costs [6]. One key challenge in federated learning is the cost of implementing the approach in software [11–13]. For example, a study by [7] developed a system called Federated Averaging (FedAvg) that reduced the communication cost of federated learning by using stochastic gradient descent (SGD) to aggregate updates from clients. Similarly, a study by [10] developed a system called MOCHA that used model compression techniques to reduce the computational cost of federated learning.

[8] propose a framework for deep learning at the edge. It aims to optimize deep learning on low-energy edge devices by architecture awareness, considering the target inference platform and introducing security and adaptiveness very early in its design. In another study, [9], this framework has been optimized for a cnn use case. The design of this framework is generic enough to also measure the amount of data transferred between the components, although the authors do not specifically mention it. We aim to provide such a mechanism to allow further evaluation of such frameworks, which should also allow us to further explain power consumption on edge devices.

Despite these efforts to optimize the cost and enhance the security of federated, and other machine learning, potential concerns still need to be addressed. For example, studies by [11,13] demonstrated that an eavesdropper could infer private data by analyzing client updates in a federated learning system. The authors showed that this attack was particularly effective when the clients had limited resources, such as memory. This highlights the effectiveness of an eavesdropper, allowing us to inspect (unencrypted) information during system execution.

These scientific works provide great additions to the field of federated learning, security and privacy-preserving machine learning, but none of them provides hard facts on how to measure security or extra overhead incurred by security measures. We aim to provide such a method, as well as a reference implementation to evaluate such overhead costs in the area of federated learning.

3   Design

In this section, we describe the design of our federated learning system and discuss the communication between clients and servers. We then introduce the design additions necessary to measure and visualize the overhead cost of federated learning step by step.

Our system consists of a central server and multiple clients, each with access to local data. An overview of the fundamental idea can be seen in Fig. 1. The goal is to train a global model using the local data on each client without the need to transfer the data to the server. We adopt the Federated Averaging algorithm proposed in [7] implemented in flwr (Federated Learning frameWoRk) [1].

We aim to detect observable costs, which include the bytes generated by each client and the time spent by each client waiting for the central server, as well as the hidden cost, which includes the time spent outside of each client by external influences, such as the network transmission, or the network topology.

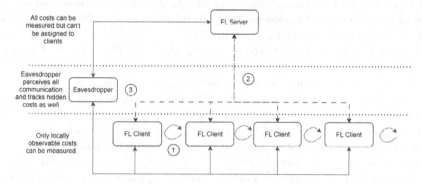

Fig. 1. Design overview of the suggested federated learning system, introducing the eavesdropper. Red marks the generation of hidden costs, green the observable costs. (Color figure online)

Each client initializes their local model, registers to the federated learning server, and starts training the first epochs, instead of synchronizing it beforehand with the federated learning server. They then continue individually training the model using its local data, which already generates bytes and thus directly observable cost, as seen in Fig. 1, item 1. Each client then sends the updated model parameters back to the server [3]. The server aggregates the updated model parameters from all clients using Federated Averaging [7] and broadcasts the updated model to all clients. This process continues for several rounds until the model converges. The difference in cost between item 1 and item 2 is that inside the clients any cost can be measured. Outside of the clients, in item 2, no client can perceive the generated cost.

The clients track the locally incurred overhead cost of the federated learning training themselves. These values include but are not limited to the size of the learning result that is to be sent to the server in bytes, the time spent calculating the size of the learning result bytes, and the time spent waiting for this client to be selected for the next training, i.e. the time spent synchronising. This local reflection cannot be seen as absolutes. For example, the learning result size in bytes in a network is just the payload of a message protocol, which itself creates other overhead. The time spent on calculating the byte size of the learning result again takes time to determine. Thus, the local overhead costs only provide an inaccurate picture of the overhead, which needs to be sharpened by another, external component.

We connect the clients to the server using GRPC [2] since this allows us to keep the components connected throughout the training rounds. Thus, the server can on the fly decide which clients to select for further training rounds and see failed or disconnected clients. But by using such a direct connection we cannot extract any meta-information about the communication, such as the actual number of transmitted bytes, or the time taken for these bytes to be received by one communication partner. To mitigate this problem we introduce the *eavesdropper*, as seen in Fig. 1, item 3.

An *eavesdropper* is a software component that acts like a network proxy. It is part of the communication network and can access the raw information transmitted from and to clients, i.e. it can listen to, or eavesdrop on, the messages transferred between the network clients. The basic premise is similar to the network security concept of a *man in the middle (mitm)* attack [5]. We introduce the *eavesdropper* as a central network component in the network. As the communication between federated learning client and server is based on the clients sending messages to the server and the server just responding, the *eavesdropper* just needs to be aware of how to reach the server and intercept all messages for it. By intercepting these messages we can, in addition to collecting the actual number of bytes being transferred over the network, including any protocol overhead like HTTP-Headers, also pinpoint the time when a message enters and leaves the network.

4 Reference Implementation

In this section, we propose a reference implementation of our federated learning system in a Kubernetes cluster environment. First, we describe the standard federated learning environment used in state-of-the-art applications, then we introduce the additions necessary to measure the overhead cost. Kubernetes is an open-source platform that automates the deployment, scaling, and management of containerized applications. Our federated learning system leverages Kubernetes to provide a scalable, fault-tolerant, and easily deployable solution for distributed machine learning.

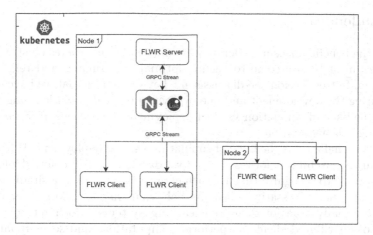

Fig. 2. Detailed overview based on our software environment. Node and pod assignments have been chosen arbitrarily and may differ during test runs.

Our system architecture, as seen in Fig. 2, consists of a central server and multiple nodes, where each node runs Kubernetes pods hosting clients. Note that Fig. 2 does not show the full cluster environment for brevity and clarity. The clients contain a containerized version of our federated learning algorithm and their local training data.

Our federated learning system also utilizes an Nginx proxy as an *eavesdropper* that listens to all communication between clients and the central server pod. The Nginx proxy is extended using Lua scripts for enhanced logging capabilities. It acts as a reverse proxy that sits between the clients and the central server pod. All client requests are routed through the Nginx proxy to the server pod, and all server responses are routed back through the proxy to the clients. By intercepting and analyzing this traffic, we can gain insights into the behaviour of the system and identify potential issues that may arise. For example, we can log the headers and payloads of requests and responses and track metrics such as request and response times. This way, we can gain a deeper understanding of how much overhead is generated in the network as a whole.

This reference implementation of a federated learning system in a Kubernetes cluster provides a scalable, fault-tolerant, and easily deployable solution for distributed machine learning. The use of Kubernetes StatefulSets, Jobs, Services and ConfigMaps enables us to manage the deployment, scaling, and configuration of our system effectively. The combination of the Nginx proxy and Lua scripts as an *eavesdropper*, along with Kubernetes for hosting, provides an efficient and scalable solution for monitoring and analyzing the traffic in our federated learning system. This allows us to not just monitor the overhead cost generated in the training services, but also track the overhead cost incurred by communication over the network.

5   Challenges

Despite the benefits of using federated learning in a Kubernetes cluster, several challenges must be overcome to ensure optimal performance and reliability of the system. In this section, we discuss some key challenges that we have encountered during the development and deployment of our federated learning system, such as the issue of simulation vs. real deployment, limitations of libraries, and optimizing synchronization.

The first challenge is the issue of simulation vs. real deployment. While simulation can be a useful tool for testing and validating a system before deployment, it is important to recognize that the behaviour of a system in a simulated environment may not necessarily reflect its behaviour in a real-world deployment. This is especially true for federated learning systems, which rely on a large number of distributed clients to perform computations and send updates to a central server. As a result, it is important to test the system thoroughly in a real-world deployment environment to ensure that it can perform optimally under real-world conditions. While we do not provide a testing environment on multiple, distributed client devices, we do deploy our reference implementation on a distributed Kubernetes cluster. This cluster itself is a distributed environment, which is closer to the real-world environment than typical federated learning simulations.

The second challenge is the limitations of libraries. While there are several libraries available for implementing federated learning systems, these libraries have certain limitations that can impact the performance of the system. For example, some libraries may not support certain types of machine learning models or may have limited support for the customization of the federated learning algorithm. In addition, some libraries may be less efficient in terms of memory usage or computation time, which can impact the scalability of the system. For *eavesdropping* there are to our knowledge no suitable software libraries, only configurable tools. While this is sufficient for our exploratory study, it could prove to be difficult to integrate the results of the *eavesdropper* from such finished tools. To address these limitations, it may be necessary to develop custom solutions or modify existing libraries to meet the specific requirements of the system.

The third challenge is optimizing synchronization. Federated learning systems rely on the synchronization of client updates to a central server to update the global model. However, the synchronization process can be a bottleneck in the system, particularly when dealing with a large number of clients. To optimize synchronization, it may be necessary to implement strategies such as batching updates, compressing updates, or using more efficient communication protocols.

The final challenge we encountered is the credibility of our measurement. As already mentioned in the previous sections, measuring possible overhead costs can also be seen as overhead, as it takes computation time away from machine learning processes. To address this challenge, we employed various techniques such as measuring the overhead of individual components separately and comparing the results with those obtained from a control group that did not use an *eavesdropper*. It is important to continually monitor and re-evaluate the measurement process as the system evolves, to ensure that the measurement results remain accurate and reliable.

6   Results

To evaluate the performance of our federated learning system in a Kubernetes cluster, we conducted experiments to measure the bytes generated by training, time spent synchronizing, and actual bytes sent over the network. The experiments were conducted using a testbed consisting of 40 federated learning clients running in Kubernetes pods. We also used a central Kubernetes pod to represent the server in the system. In this testbed, we train the clients using the LEAF FEMNIST dataset [4], 100 server epochs, i.E. the server triggers training and evaluation 100 times on selected clients. We evaluate these clients by the bytes they generate during the training, the time spent waiting on synchronisation and selecting and the actual bytes sent over the network we perceive by employing an *eavesdropper*. Further, we also evaluate the impact of the *eavesdropper* in the federated learning system to create a hypothesis on the scalability of this approach.

In the following subsections, we discuss the results shown in Table 1 in terms of bytes generated and sent over the network and the time spent synchronising the clients, before finally discussing the impact of introducing an *eavesdropper* to the system, which is also shown in Table 1. We sampled ten random clients for evaluation in these tables for brevity.

6.1   Bytes Sent

The size of bytes sent over the network is directly proportional to the data used to train the network. Clients 2 and 4 for example have a more significant divergence in training data sizes. Client 2 uses ∼ 3000 samples, while Client 4 uses about twice as many samples, which is reflected in the size of the transmitted data. Surprisingly, the structure of the network, while also relevant for the size of the transmitted data, does not impact the overall size as much as the amount of training data.

Most notable is the divergence in the number of bytes sent and the number of bytes we have detected with an *eavesdropper* in place. As seen in Table 1, the *eavesdropper* makes the communication size visible. Local measurements do not include the overhead of the message protocols and networking procedures, which is the decisive overhead in comparison. We attribute this divergence in size to the selected communication protocol, GRPC fork-join streams.

6.2   Synchronisation Time

The larger the training data, the
longer one client needs to train
an epoch. In flwr [1], a finished
client that already transmitted its
results is idling and waiting for a
new selection round. We define this
time as part of the synchronisation time, hence we can assume the
synchronisation time per client to
be indirectly proportional to the
size of the training data. As seen
in Fig. 3, clients with few training samples, such as client 2, fin-
ish their training faster, resulting
in a longer idle time while they
wait for clients with larger training data sets to complete, such as client 5.

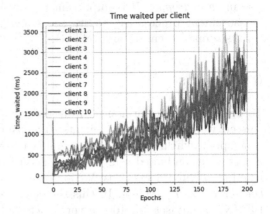

Fig. 3. Time waited per client as per one test run.

Comparing the average time spent synchronization with and without an
eavesdropper did not result in conclusive results. No major outliers were identified with 40 individual clients, all training individually on different training
data sizes. The major discrepancies here can be traced back to random CPU
allocations of the Kubernetes cluster.

Table 1. Performance Metrics for Federated Learning Clients, with *eavesdropping*.

Client ID	Total Bytes Sent	Avg. Time Synced	Avg. Bytes Sent	*Eavesdropped Bytes*	*Eavesdropped Avg. Time Synced*
1	3.41 MB	12.70 s	17.14 KB	13.87 MB	11.90 s
2	2.08 MB	13.44 s	10.46 KB	13.24 MB	14.45 s
3	4.16 MB	8.49 s	21.02 KB	13.73 MB	7.53
4	4.16 MB	8.60 s	21.02 KB	14.07 MB	8.32 s
5	4.70 MB	5.72 s	23.77 KB	14.15 MB	7.10 s
6	4.10 MB	7.21 s	20.74 KB	13.78 MB	9.23 s
7	3.52 MB	9.89 s	17.77 KB	13.74 MB	9.76 s
8	2.07 MB	13.51 s	10.50 KB	13.86 MB	13.05 s
9	3.40 MB	9.19 s	17.20 KB	13.87 MB	9.94 s
10	2.07 MB	13.98 s	10.50 KB	13.87 MB	12.80 s

6.3   The Cost of Eavesdropping

The *eavesdropper* itself does need resources we can track, such as CPU and memory usage. We detected some minor peaks on larger loads, but nothing uncom-
mon when comparing it to state-of-the-art reverse proxies used in modern software systems. We did find potentially larger costs for the system while inspecting
the memory usage of the *eavesdropper*. The RAM necessary to properly track the
federated learning clients without loss or noise is directly proportional to both

the number of clients and the number of epochs they train. So we can formulate the total memory necessary, just for the *eavesdropper* in MB as $TotalMemory = Clients \times Epochs + k$, where k is a constant representing the initial ram usage of the *eavesdropper*. This formula was derived based on the observation of memory usage in a specific set of scenarios involving different numbers of clients and epochs. While this formula may provide a useful estimate for similar scenarios, it may not accurately predict memory usage in all cases. Other factors, such as the size and complexity of the input data and the implementation of the *eavesdropper* may also have an impact on memory usage.

This behaviour can be explained by the communication protocol we use in our reference implementation, as the connection between the federated clients and the server will not close until the federated model has finished training. This could lead to performance issues when scaling the reference implementation to even larger loads, which would independently occur on the federated learning server, even without the *eavesdropper*.

7 Conclusion and Future Work

In conclusion, this paper presents a detailed examination of the cost of federated learning using a reference implementation of federated learning on a Kubernetes cluster and an Nginx proxy with Lua scripts for eavesdropping. The implementation was evaluated using the LEAF FEMNIST dataset and performance metrics such as bytes sent and time spent synchronizing was measured. The results show that the system is effective and scalable, with good performance even with a large number of clients. However, the *eavesdropper*'s RAM usage was found to be a potential drawback, as it increased significantly with the number of clients and epochs in the system.

Overall, the reference implementation presented in this paper provides a useful starting point for those interested in exploring the overhead cost of their federated learning environments. The performance metrics demonstrate the system's effectiveness and scalability, while the identified drawbacks can help guide future improvements to the system. In particular, addressing the issue of RAM usage for *eavesdroppers* will be an important area of future work.

Acknowledgements. S3AI is a COMET Module within the COMET - Competence Centers for Excellent Technologies Programme and funded by BMK, BMAW and the State of Upper Austria. The COMET Programme is managed by FFG. The research reported in this paper has been funded by the Federal Ministry for Climate Action, Environment, Energy, Mobility, Innovation and Technology (BMK), the Federal Ministry for Labour and Economy (BMAW), and the State of Upper Austria in the frame of the COMET Module Security and Safety for Shared Artificial Intelligence by Deep Model Design (S3AI) [(FFG grant no. 872172) and the SCCH competence center INTE-GRATE [(FFG grant no. 892418)] within the COMET - Competence Centers for Excellent Technologies Programme managed by Austrian Research Promotion Agency FFG.

References

1. FLWR: a federated learning framework. https://flwr.dev/ (2023). Accessed 23 Feb 2023
2. gRPC (2023). https://grpc.io/. Accessed 6 June 2023
3. Bonawitz, K., et al.: Towards federated learning at scale: system design. Proceed. Mach. Learn. Syst. **1**, 374–388 (2019)
4. Caldas, S., et al.: LEAF: a benchmark for federated settings (2018). arXiv:1812.01097
5. Callegati, F., Cerroni, W., Ramilli, M.: Man-in-the-middle attack to the https protocol. IEEE Secur. Priv. **7**(1), 78–81 (2009)
6. Konečný, J., McMahan, H.B., Yu, F.X., Richtárik, P., Suresh, A.T., Bacon, D.: Federated learning: strategies for improving communication efficiency. arXiv preprint arXiv:1610.05492 (2016)
7. McMahan, B., Moore, E., Ramage, D., Hampson, S., Arcas, B.A.: Communication-efficient learning of deep networks from decentralized data. In: Artificial Intelligence and Statistics, pp. 1273–1282. PMLR (2017)
8. Meloni, P., et al.: ALOHA: an architectural-aware framework for deep learning at the edge. In: Martina, M., Fornaciari, W. (eds.) Proceedings of the Workshop on Intelligent Embedded Systems Architectures and Applications, INTESA@ESWEEK 2018, Turin, Italy, 04 October 2018, pp. 19–26. ACM (2018). https://doi.org/10.1145/3285017.3285019
9. Meloni, P., et al.: Optimization and deployment of CNNs at the edge: the ALOHA experience. In: Palumbo, F., Becchi, M., Schulz, M., Sato, K. (eds.) Proceedings of the 16th ACM International Conference on Computing Frontiers, CF 2019, Alghero, Italy, 30 April - 2 May 2019, pp. 326–332. ACM (2019). https://doi.org/10.1145/3310273.3323435
10. Smith, V., Chiang, C.K., Sanjabi, M., Talwalkar, A.S.: Federated multi-task learning. In: Guyon, I., et al. (eds.) Advances in Neural Information Processing Systems, vol. 30. Curran Associates, Inc. (2017). https://proceedings.neurips.cc/paper_files/paper/2017/file/6211080fa89981f66b1a0c9d55c61d0f-Paper.pdf
11. Wang, L., Xu, S., Wang, X., Zhu, Q.: Eavesdrop the composition proportion of training labels in federated learning (2019)
12. Zellinger, W., et al.: Beyond federated learning: on confidentiality-critical machine learning applications in industry. Procedia Comput. Sci. **180**, 734–743 (2021). https://doi.org/10.1016/j.procs.2021.01.296. https://www.sciencedirect.com/science/article/pii/S1877050921003458. proceedings of the 2nd International Conference on Industry 4.0 and Smart Manufacturing (ISM 2020)
13. Zou, Y., Wang, G.: Intercept behavior analysis of industrial wireless sensor networks in the presence of eavesdropping attack. IEEE Trans. Industr. Inf. **12**(2), 780–787 (2015)

Cyber-Security and Functional Safety in Cyber-Physical Systems

A Context Ontology-Based Model to Mitigate Root Causes of Uncertainty in Cyber-Physical Systems

Mah Noor Asmat[1]([✉])[iD], Saif Ur Rehman Khan[2][iD], Atif Mashkoor[3][iD], and Irum Inayat[4][iD]

[1] Department of Computer Science, COMSATS University Islamabad (CUI), Islamabad, Pakistan
mahnoorasmat18@gmail.com
[2] Department of Computing, Shifa Tameer-e-Millat University (STMU), Islamabad, Pakistan
saif_rehman.ssc@stmu.edu.pk
[3] Institute for Software Systems Engineering, Johannes Kepler University (JKU), Linz, Austria
atif.mashkoor@jku.at
[4] Department of Computer Science, National University of Computer and Emerging Sciences (FAST-NUCES), Islamabad, Pakistan
irum.inayat@nu.edu.pk

Abstract. A Cyber-Physical System (CPS) is a networked collection of diverse physical elements that perform complex operations to achieve a certain goal. To ensure the quality and reliability of a CPS, uncertainty is regarded as one of the crucial challenges that need to be effectively handled. However, the current state-of-the-art lacks in focusing on handling the root causes of uncertainty in the context of CPS. This study proposes a Context Ontology-based Uncertainty Mitigation (COUM) model to mitigate uncertainty during the early phases of CPS's software development life cycle like requirement elicitation. The proposed COUM model intends to identify and mitigate the root causes of uncertainty to improve the dependability of CPS. The COUM model is applied to the Care-o-Bot system to address its uncertainties and increase its reliability.

Keywords: Cyber-Physical Systems · Uncertainty · COUM Model

1 Introduction

Cyber-Physical Systems (CPS) are used in every aspect of our daily lives, such as robots, implantable medical devices, intelligent buildings, self-driving cars, and planes that fly in controlled airspace. A CPS is a set of heterogeneous physical elements that take physical values from the actual world and execute complicated operations to accomplish a goal [1,2]. As a result of multi-disciplinary interaction, the complexity of CPS significantly increases, posing many issues

© The Author(s), under exclusive license to Springer Nature Switzerland AG 2023
G. Kotsis et al. (Eds.): DEXA 2023 Workshops, CCIS 1872, pp. 45–56, 2023.
https://doi.org/10.1007/978-3-031-39689-2_5

such as security, uncertainty, robustness, architectural issues, and privacy [3,4]. Security and uncertainty are two significant concerns for the CPS [5–7].

When discussing CPS and its dependability, the term "uncertainty" refers to a lack of knowledge or understanding regarding a specific system's circumstances, environmental changes, or the kind of data or outcome that could affect the system's dependability, availability, safety, or other attributes. It includes the ambiguity or unpredictability related to various factors, including system behavior, outside influences, and interactions between system components, which may result in risks or vulnerabilities for the CPS [1]. For example, if there is a healthcare robot that helps an elderly person in an emergency condition. In this case, Remote Operators (RO) must understand and comprehend the emergency to help an elder person. However, in this case, the robot does not correctly comprehend the situation and cannot provide the necessary communication to handle the emergency. Indeed, it occurs mainly due to uncertainty in the robot.

Uncertainty can be classified into two categories: (i) known and (ii) unknown. Known uncertainty refers to a situation in which probability is precisely determined. Conversely, the person is uninformed about the likelihood of unknown uncertainty. However, the person believes someone else may know it. In the literature, the causes of uncertainty are discussed in various ways. Modeling and simulation can generally reduce and control known uncertainties, while unknown uncertainties can only be identified and addressed during runtime [1]. Furthermore, in distinct application contexts of CPS, such as production systems, medical systems, power grids, and autonomous vehicles, uncertainty exists in different parts. It is crucial to understand how uncertainty affects CPS. Furthermore, as the complexity of the CPS grows, so does uncertainty. Therefore, uncertainty must be mitigated for the CPS to work with reliability.

In the early stages of a software project, occurrence chance or probability is known for the known uncertainty. In contrast, the occurrence chance or probability of the unknown uncertainty is unknown until its occurrence [8]. The scenario determines the unknown parts of uncertainty in CPS, the event, the timing and/or probability, the source, and so on. Depending on the context, it could be any of these aspects. Modeling and testing reduce and mitigate known uncertainty, whereas unknown uncertainty is only detected at runtime [1]. From the initial stage of development, uncertainty modeling is used to deal with known uncertainty in CPS.

This work discusses uncertainty handling approaches from the current state-of-the-art literature. Furthermore, the root causes of uncertainty are identified from the literature. As a result, a Context Ontology-based Uncertainty Mitigation (COUM) model is proposed based on context ontology to mitigate the root causes of uncertainty in CPS. A context ontology defines a common vocabulary to express context information. It also contains machine-interpretable definitions of the domain's fundamental concepts and relationships.

The main contributions of this study are as follows:

- Proposes a meta-model of Context Ontology-based Uncertainty Mitigation (COUM) model to introduce the relationship between environment and uncertainty.
- Develops a COUM model for mitigation of root causes of uncertainty.

The remaining parts of this paper are organized as follows: Sect. 2 provides the literature review. The COUM model is presented in Sect. 3, and applied to a case study in Sect. 4. Results and discussions are provided in Sect. 5. Finally, Sect. 6 concludes the current work and highlights potential future research directions for interested researchers.

2   Literature Review

In the literature, few studies focused on handling the uncertainty by implementing current state-of-the-art approaches are discussed [9–12]. The reported studies primarily address uncertainty in the context of CPS. However, a related field of research is known as system-of-systems engineering, which also deals with similar challenges. System-of-systems engineering involves integrating multiple independent systems into a larger, interconnected system to achieve specific goals. It emphasizes the inter-dependencies, emergent behavior, and uncertainty arising from these individual systems' interactions. Uncertainty and a causal relationship are modeled through the Bayesian network [9]. Automated testing is performed through a framework, the "Uncertainty Modeling Framework" (UMF) is used to tackle uncertainty by creating test-ready models of uncertainty [5]. Moreover, an uncertainty model named UncerTest [10] is developed to generate the test cases and minimize the test cases of a CPS. The test cases are generated through the belief state machine. Another study [11] developed a holistic test case specification method and implemented the proposed method on the co-simulation experimental setup. In contrast, Palensky et al. [12] concluded that object-oriented modeling approaches are preferred to increase readability and interpretability.

Qasim et al. [13] emphasized the necessity of data identification and tool development in their discussion of the significance of system reconfiguration in complex systems management. They outlined the idea of an ontology, which was intended to formalize the process of system reconfiguration and address the multi-domain nature and variety of system types. The study also stated that the proposed ontology was created using expert knowledge and real-world use cases and tested using an industrial case study. Furthermore, Yang et al. [14] provided a review to assess the influence of ontologies in systems engineering knowledge areas and offer a comprehensive summary of the current state of ontology-based systems engineering. It addressed vital aspects such as systems engineering knowledge areas supported by ontologies, the contribution of ontologies to systems engineering problems, existing ontologies in support of systems engineering, and techniques adopted from an ontology engineering perspective.

Conversely, Daun et al. [15] demonstrated the importance of a systematic modeling method for documenting the overlapping contexts of concurrently engineered, collaborating cyber-physical systems. The authors concluded that when systems are aggregated into a common super-system, it is vital to systematically deduce syntactic consistency and functional efficiency. An ontology-centric context framework is presented to accomplish syntactic consistency and functional efficiency. They mentioned how to delineate the context of a system under development and how context objects such as knowledge sources, context functions, or context entities can be documented. Furthermore, dedicated context models can document externally visible states of context objects and their respective impacts on the system under development. Their outstanding uses in a running adaptive cruise control example are discussed.

Based on the current state-of-the-art, it can be concluded that the impact of uncertainty should be tackled according to its dimensions during the CPS development process to avoid runtime challenges and rework. Uncertainty is a challenge that occurs during development and even during runtime. Thus, studying the approaches and tools used to handle uncertainty in CPS is essential. This work proposes a COUM model to mitigate the root causes of uncertainty in CPS.

3   Context Ontology-Based Uncertainty Mitigation (COUM) Model

This section presents the Context Ontology-based Uncertainty Mitigation (COUM) model useful for effectively mitigating the root causes of uncertainty. The benefits of the proposed COUM model are listed as follows:

- It encourages the systematic engineering of numerous simultaneously developed, interrelated components by mitigating the root causes of uncertainty.
- It provides the basis for identifying and mitigating the root causes of uncertainty during requirement elicitation and system integration.

3.1   Justification

Context theory necessitates separating the system under development from the objects in its environment. Specifically, the COUM model ontology permits differentiating system components under development from aspects of the context that impact the system. However, it cannot be modified by the developer [16]. The context objects may include people (or other biological entities), generic information sources, and external systems (software). Context objects interact with the context subject after deployment, for example, users or other systems, or give knowledge about the subject or its development, such as physical constraints or laws [17]. As a result, ontology distinguishes between two distinct primary forms of context. The context might be determined as the system's knowledge context or as the system's operational context. Moreover, the context of knowledge includes all sources of information that limit the system's evolution [18]. Conversely, the operational context concentrates on the interaction between the context subject and context objects [15,19]. Furthermore, uncertainty occurs in the environment and context where the system is deployed. To mitigate uncertainty in CPS, it is essential to identify and mitigate the root causes of uncertainty in the CPS context [20].

3.2 Proposed Meta-Model

To mitigate the root causes of uncertainty, the context ontology framework [19] is integrated with the uncertainty ontology model [20]. Furthermore, the root causes of uncertainty must also be documented in the CPS context. A context ontology, part of an environment, is depicted in the meta-model in Fig. 1. While uncertainty is associated with the context within an environment, uncertainty occurs in the context of CPS, which consists of specific root causes. The context ontology further consists of the context of knowledge and the operational context that interact with the context subject [16]. The colored elements of the model depict the components presented in the current work.

Environment. The environment is a component that can consist of uncertainty. Context is part of an environment. Changes in the environment impact the system in context. The impact of the uncertainty on CPS and its context no longer depends on the individual system. Instead, it depends on the functional interaction that arises at runtime within the system when it communicates and interacts. This implies that uncertainty due to functional dependency occurs not only within a single CPS but also between several of them within the environment. The highlighted parts of the meta-model illustrate an environment, as

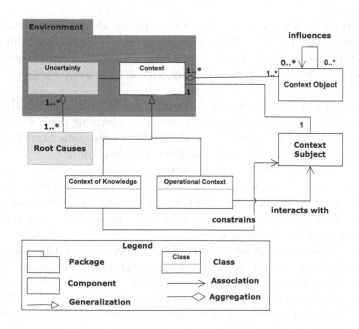

Fig. 1. A Meta-Model of the COUM Model.

shown in Fig. 1. An environment consists of uncertainty and a CPS context. A CPS context lies within an environment that consists of the contexts of knowledge and operational context. The context of knowledge constrains the context subject. Understanding the relationship between environment, uncertainty, and a CPS context is important.

Root Causes. Uncertainty consists of root causes in the context, and uncertainty lies within an environment. Root causes of uncertainty exist in three types of the domain in a CPS context; natural process, human behavior, and technological process as depicted in our previous studies [21,22]. It is required to identify uncertainty and its root causes in a CPS context and mitigate them during the elicitation of requirements through the COUM model.

3.3   COUM Model

The COUM model is a proposed approach for mitigating uncertainties in CPS during the early stages of software development. It aims to identify the root causes of uncertainty and provide strategies to mitigate them. The COUM model utilizes a context ontology, which defines a common vocabulary to express contextual information and machine-interpretable definitions of fundamental concepts and relationships within the domain. The proposed COUM model is shown in Fig. 2. The COUM model consists of several key components and processes:

1. Context Ontology: The model starts by developing a context ontology, representing the contextual knowledge required to understand the CPS and its environment. The ontology defines the context objects, context subject, and their relationships.

2. Uncertainty Identification: The COUM model identifies different types of uncertainties within the CPS. This includes uncertainties arising from environmental factors, system-related and human interactions. A comprehensive understanding of the challenges is obtained by considering these various sources of uncertainty.

3. Root Cause Analysis: Once uncertainties are identified, the COUM model facilitates a root cause analysis to determine the underlying factors contributing to the uncertainties. This involves examining the interactions between the CPS and its environment and the context objects and their impacts.

4. Uncertainty Mitigation Strategies: Based on the root cause analysis, the COUM model proposes mitigation strategies to address the identified uncertainties. These strategies may involve refining the system design, incorporating additional sensing or monitoring capabilities, enhancing communication protocols, or introducing redundancy measures.

The COUM model emphasizes integrating uncertainty mitigation strategies into the requirement elicitation and system integration processes. Considering uncertainties from the early stages of development, potential challenges and risks can be addressed proactively, leading to more reliable and robust CPS. The COUM model aims to provide a systematic approach to handling uncertainties in CPS by leveraging a context ontology and considering various sources of uncertainty.

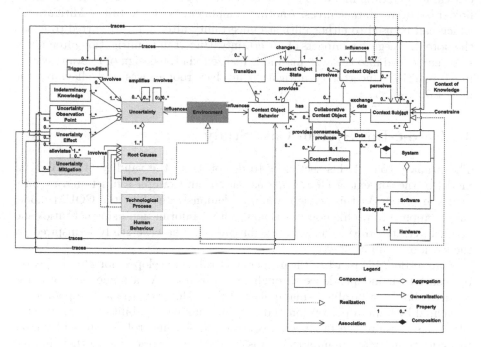

Fig. 2. A Context Ontology-based Uncertainty Mitigation (COUM) Model in CPS.

Uncertainty mitigation depends on systematically identifying potential root causes of uncertainty in its environment and collecting evidence that uncertainties are unlikely to occur during runtime. Knowledge sources that help the CPS develop are included in the context of knowledge. These knowledge sources could be engineered artifacts from the contextual subject or other systems to a large extent. In this way, the context of knowledge is transformed into a library of engineering artifacts explicitly created for the "context subject." We have constrained the ontologies to artifacts that directly explain the requirements of the context subject and context objects.

The environment influences the uncertainty and "context object behavior" in a CPS context. Transitions occur in the "context object behavior" due to a trigger condition that involves uncertainty. The trigger condition can indicate changes in the behavior of "context objects" as well as the data collected as a result of those changes. Note that uncertainty does not need to have a trigger condition; it could also exist naturally. Therefore, the uncertainty effect refers to the impact of uncertainty on the system or its context. The result could be unfavorable and lead to harmful conduct.

At runtime, the system is aware of changes in its operational context and uncertainty caused by information such as sensor data, object flows, messages, or human interaction. The notion of "Observation Point" describes which artifacts contain uncertainty and where the system may identify that uncertainty. This is critical for determining how the system is aware of it. An observation site can be linked to several data sources. Another component is "Uncertainty Mitigation," which can be used to either mitigate uncertainty after it has occurred to lessen the potential negative impacts of uncertainty. Uncertainty mitigation often needs to be embedded into the systems and is realized via the design of relevant system functionalities. Uncertainty mitigation involves root causes and alleviates the uncertainty effect.

4   Care-o-Bot System Case Study

The primary goal of this section is to demonstrate the validity of the COUM model in the context of CPS. For this reason, an example's functional requirements are given. In this section, we have demonstrated how the COUM model can be applied to profile non-functional context information using activities and artifacts created during uncertainty mitigation. The case study is inspired by the work of Daun et al. [17].

Care-o-Bot [17] is a healthcare personal robot developed for elderly people to ensure their independence through service robots. As a home caregiver, the service robot must be able to manipulate objects. However, robot manipulation is prevented by limited perception and decision-making capabilities in unexpected situations. Despite the progress in control and sensing, robots are still unable to perform uncertain manipulation tasks in the real world due to their limited perception and decision-making abilities.

The Shadow Robotic System (SRS) is used in this scenario to assess the severity of an emergency, such as a fall. After pressing the wearable alarm button, the robot approaches the user, allowing the RO to get a clear view of the matter and attempt an initial communication with the user in distress. This initial interaction will enable him to comprehend the current situation better and take appropriate action.

Suppose the user recovers quickly from the fall and communicates that an ambulance is unnecessary. In that case, the RO may decide that dispatching emergency response services is unnecessary or that they are dispatched to check on the elderly person as a low-priority service. On the other hand, if the user cannot recover from the fall and/or does not interact with any efforts at communication, the case's response level can be raised, and an ambulance can be dispatched as soon as possible. In contrast to the "emergency button" alone service, if the user is unable to recover from the fall and/or does not interact with the SRS system allows the RO to gain a better understanding of the situation because, unlike a smart home or "emergency button" support operator, the robot can move around to obtain a better-unobstructed view, open doors when necessary, and fetch objects for the user. Any attempts at communication will escalate the case's response level, and an ambulance will be dispatched as soon as possible.

Suppose emergency response services, such as an ambulance, are called to the older person's home. In that case, the robot moves close to the front door in a safe position, not blocking it in the event of a malfunction, and opens the door from the inside with the RO's permission and, if necessary, assistance. In this circumstance, video observation via the robot's cameras is critical once the emergency button has been touched, so it is not an optional assignment. Suppose the elderly person's condition necessitates regular checks. In that case, it is feasible to schedule a periodic operation for the robot to come and check on the user's condition, provided that this has been agreed upon in advance. For example, the robot may make brief trips to the user every hour to allow the RO to assess the situation, such as whether the user has collapsed. The functional requirements of the Care-o-Bot system are discussed as follows:

1. UI Functions
   - UI is a human-robot interaction that connects the robot and the RO regarding perception and decision-making. It motivates and intervenes with the robot while providing feedback to the RO. A RO's motivation allows them to provide the robot with goals. Depending on its motivations, such as uncertainty mitigation, the robot may choose to accomplish some or all of these goals. Through intervention, autonomous robotic perception and decision-making are improved. Furthermore, feedback allows RO to assess the outcome of the intended task activity. However, perception and other capabilities (navigation) are only used to validate this work.
2. Perception
   - PER_F1 Known Object Recognition: Recognize a learned object for manipulation. Most SRS manipulation operations require this capability, which is also increased by user demand: R08 Objects (shapes, colors, letters on food packaging, and numbers on microwave display) are recognized and identified by the system.
   - PER_F2 Environment Perception for Navigation: Perceive and comprehend the environment to complete navigation tasks.
   - PER_F3 Object Manipulation Environment Perception: Perceive and interpret the environment for manipulative tasks. The SRS manipulation and navigation require the PER_F2 and PER_F3 capabilities. It was also required for the following user needs: R05 The system recognizes various housing environments. R06 The system navigates by identifying and preventing obstacles.
   - PER_F4 Learning New Objects: During the operation, learn new objects. SRS requires this ability to grasp unexpected items. It was also required for the following user need: R08 Objects are recognized and identified by the system.
   - PER_F5 Human Motion Analysis and Tracking: identify and track users in the environment. The SRS robot can use this function to watch the human's location, pose, and actions to deduce information about the user's movements, gestures, and intentions. A user demand also prompts it: R04 user position in the environment is recognized and traced.

3. Other Functions
  - NAV_MAN1 Navigation/Manipulation Action by User: The strategy comprises the target and context information.
  - NAV_MAN2 Navigation/Manipulation by Direct Motion Instruction: Perform the navigating/manipulating motion using direct instruction. These two functions connect the SRS control system to the hardware platform of choice.

In the example of a healthcare robot, the robot is influenced by the behavior of the surrounding environment where it works. Sensor failure in an emergency may change the robot's interpretability. The robot may be unable to move due to uncertainty occurred. To handle the uncertainties occurred on runtime, uncertainty needs to be identified and mitigated during requirement elicitation. Engineers gather requirements during requirement elicitation. Experts divide those requirements into two types. Requirements with uncertainty and requirements without uncertainty. If the experts identify the uncertainty in requirements. Uncertain requirements are further processed through the identification of the root causes. Consequently, possible solutions to the root causes are identified by experts. Then, root causes are mapped in the CPS context where uncertainty occurs. To mitigate the root causes of uncertainty, the COUM model maps the root causes in a CPS context. The components to perform mapping is illustrated in the COUM model in Fig. 2. Each root cause of uncertainty must have certain action steps to mitigate them.

5   Results and Discussion

To apply the COUM model to the Care-o-Bot system, it was necessary to address the risks and difficulties associated with healthcare robotics. The objective was to show that the COUM model effectively reduces the underlying sources of uncertainty and enhances CPS dependability.

In the case study, the Care-o-Bot system was created as a personal healthcare robot for elderly people to help them maintain their independence. The system had issues with robot manipulation, limited perception, and unexpected situational decision-making. The COUM model was used to profile non-functional context information and reduce uncertainties to address these issues. The case study concentrated on emergency response and user interaction scenarios and was motivated by the work of Daun et al. [17].

The Care-o-Bot system uses the Shadow Robotic System (SRS) to evaluate the seriousness of emergencies like falls. The robot would approach the user, ascertain the circumstances, and then decide what to do in response to the user's situation and communication. The system's goal was to efficiently dispatch emergency response services while considering the user's cure and interaction abilities.

Human-robot interaction, perception capabilities for object recognition and environment understanding, learning new objects, and human motion analysis and tracks were covered in the functional requirements of the Care-o-Bot system. These requirements were essential to the system's ability to function properly.

The COUM model was used in handling uncertainty to pinpoint and address its underlying causes during requirement elicitation. To identify the underlying causes, uncertain requirements were discovered and further processed. The root causes of uncertainty in the CPS context were mapped, and potential solutions were identified. In general, the COUM model's application to the Care-o-Bot system showed how the model could be used to effectively address uncertainties and increase the CPS's reliability in the healthcare robotics field. The COUM model diagram's components served as guidance for identifying the underlying causes and formulating a plan of action to reduce uncertainty.

6 Conclusion and Future Directions

This paper provides the current state-of-the-art uncertainty handling approaches and root causes of uncertainty in the CPS domain. Moreover, a Context Ontology-based Uncertainty Mitigation (COUM) model is proposed to identify and address underlying root causes of uncertainty in the CPS context. A case study of Care-o-Bot is adapted from the literature to validate the COUM model. Care-o-Bot is an assisting robot to help elderly people in emergency conditions. Some of the functionalities of the adopted case study are implemented in the COUM model. The model diagram's components served as guidance for identifying the underlying causes and formulating a plan of action to reduce uncertainty. Based on the attained results, it is concluded that the proposed COUM model can effectively mitigate uncertainty by identifying and mitigating the root causes.

In the future, we plan to implement the COUM model on an industrial CPS to mitigate uncertainty. Moreover, we plan to conduct further case studies to compute the scalability of the COUM model for mitigating the uncertainty in the CPS context.

References

1. Ali, S., Yue, T.: U-test: evolving, modelling and testing realistic uncertain behaviours of cyber-physical systems. In: 2015 IEEE 8th International Conference on Software Testing, Verification and Validation (ICST), pp. 1–2. IEEE (2015)
2. Delicato, F.C., Al-Anbuky, A., Wang, K.I.-K.: Smart cyber-physical systems: toward pervasive intelligence systems. Future Gener. Comput. Syst. **107**, 1134–1139 (2020)
3. Esterle, L., Grosu, R.: e & i Elektrotechnik und Informationstechnik **133**(7), 299–303 (2016). https://doi.org/10.1007/s00502-016-0426-6
4. Ong, L.M.T., Nguyen, N.T., Luong, H.H., Tran, N.C., Huynh, H.X.: Cyber physical system: Achievements and challenges. In: Proceedings of the 4th International Conference on Machine Learning and Soft Computing, pp. 129–133 (2020)

5. Zhang, M., Selic, B., Ali, S., Yue, T., Okariz, O., Norgren, R.: Understanding uncertainty in cyber-physical systems: a conceptual model. In: Wasowski, A., Lonn, H. (eds.) ECMFA 2016. LNCS, vol. 9764, pp. 247–264. Springer, Cham (2016). https://doi.org/10.1007/978-3-319-42061-5_16
6. Mashkoor, A., Johannes, S., Miklós, B., Egyed, A.: Security-and safety-critical cyber-physical systems. J. Softw. Evolut. Process 32(2), e2239 (2020)
7. Biro, M., Mashkoor, A., Sametinger, J., Seker, R.: Software safety and security risk mitigation in cyber-physical systems. IEEE Softw. 35(1), 24–29 (2018)
8. Knight, F.H.: Risk, uncertainty and profit. New York: Houghton Mifflin 31 (1921)
9. Fenton, N., Krause, P., Neil, M.: Software measurement: uncertainty and causal modeling. IEEE Softw. 19(4), 116–122 (2002)
10. Zhang, M., Ali, S., Yue, T.: Uncertainty-wise test case generation and minimization for cyber-physical systems. J. Syst. Softw. 153, 1–21 (2019)
11. Van der Meer, A.A., et al.: Cyberphysical energy systems modeling, test specification, and co-simulation based testing. In: 2017 Workshop on Modeling and Simulation of Cyber-Physical Energy Systems (MSCPES), pp. 1–9. IEEE (2017)
12. Palensky, P., Widl, E., Elsheikh, A.: Simulating cyber- physical energy systems: challenges, tools and methods. IEEE Trans. Syst. Man Cybern. Syst. 44(3), 318–326 (2013)
13. Qasim, L., Hein, A.M., Olaru, S., Garnier, J.L., Jankovic, M.: System Reconfiguration Ontology to Support Model-based Systems Engineering: approach Linking Design and Operations. Systems Engineering, Wiley Online Library (2023)
14. Yang, L., Cormican, K., Yu, M.: Ontology-based systems engineering: a state-of-the-art review. Comput. Ind. 111, 148–171 (2019)
15. Daun, M., Brings, J., Weyer, T., Tenbergen, B.: Fostering concurrent engineering of cyber-physical systems a proposal for an ontological context framework. In: 2016 3rd International Workshop on Emerging Ideas and Trends in Engineering of Cyber-Physical Systems (EITEC), pp. 5–10. IEEE (2016)
16. Jackson, M.: The world and the machine. In: 1995 17th International Conference on Software Engineering, p. 283. IEEE (1995)
17. Daun, M., Tenbergen, B.: Context modeling for cyber-physical systems. J. Softw. Evol. Process. 35, e2451 (2022)
18. Daun, M., Brings, J., Tenbergen, B., Weyer, T.: On the model-based documentation of knowledge sources in the engineering of embedded systems. In: Gemeinsamer Tagungsband der Workshops der Tagung Software Engineering 2014, pp. 67–76. CEUR-WS.org (2014)
19. Pohl, K., Broy, M., Daembkes, H., Hönninger, H.: SPES XT context modeling framework. In: Advanced Model-Based Engineering of Embedded Systems, pp. 43–57. Springer, Cham (2016). https://doi.org/10.1007/978-3-319-48003-9_4
20. Bandyszak, T., Daun, M., Tenbergen, B., Kuhs, P., Wolf, S., Weyer, T.: Orthogonal uncertainty modeling in the engineering of cyber-physical systems. IEEE Trans. Autom. Sci. Eng. 17(3), 1250–1265 (2020)
21. Asmat, M.N., Khan, S.U.R., Hussain, S.: Uncertainty handling in cyber-physical systems: State-of-the-art approaches, tools, causes, and future directions. J. Softw. Evolut. Process 35, e2428 (2022)
22. Asmat, M.N., Khan, S.U.R., Mashkoor, A.: A conceptual model for mitigation of root causes of uncertainty in cyber-physical systems. In: Kotsis, G., et al. (eds.) DEXA 2021. CCIS, vol. 1479, pp. 9–17. Springer, Cham (2021). https://doi.org/10.1007/978-3-030-87101-7_2

Architecture for Self-protective Medical Cyber-Physical Systems

Michael Riegler[1]($\boxtimes$) (iD), Johannes Sametinger[1] (iD), and Jerzy W. Rozenblit[2] (iD)

[1] LIT Secure and Correct Systems Lab and Institute of Business Informatics,
Johannes Kepler University Linz, Linz, Austria
{michael.riegler,johannes.sametinger}@jku.at
[2] Department of Electrical and Computer Engineering and Department of Surgery,
University of Arizona, Tucson, USA
jerzyr@arizona.edu
https://www.jku.at/en/lit-secure-and-correct-systems-lab,
https://www.se.jku.at, https://ece.arizona.edu

Abstract. The *Internet of Medical Things* (IoMT) promises to improve patient care and the efficiency of *Medical Cyber-Physical Systems* (MCPSs). At the same time, the connectivity increases the security risk. We aim to model *Self-protective MCPSs* to reduce the attack surface during runtime. Even under attack, these systems require to provide clinical function for the patients. Monitoring vulnerabilities and suspicious behavior and sharing attacker information contributes to improved security and can be the foundation for automated actions for healthcare delivery organizations. Switching between context-aware security modes provides a flexible way to protect online and offline IoMT and increase patient safety. This paper presents our ongoing work to make healthcare systems more secure. We show current security and privacy challenges, discuss how self-protective systems can overcome them, and what role IoMT devices play in that context.

Keywords: Self-Protection · Medical Cyber-Physical Systems · Internet of Medical Things · Security · Mode Switching

1 Introduction

COVID-19 has pushed the development and usage of *Internet of Medical Things* (IoMT) devices. In critical times of lockdowns, such interconnected medical devices combined with medical sensors, actuators, applications, and services allow remote medical care delivery and reduce in-person visits. IoMT includes wearable medical devices to monitor blood pressure, heart rate, glucose, and other measures for chronic diseases, medicine pumps, fall detection sensors, remotely accessible medical implants, and connected clinic and hospital devices up to remote robotic surgical assistants. According to Statista [29], the global market value of IoMT will reach over 260 billion US dollars in 2027. However,

G. Kotsis et al. (Eds.): DEXA 2023 Workshops, CCIS 1872, pp. 57–66, 2023.
https://doi.org/10.1007/978-3-031-39689-2_6

these advantages in technology and connectivity are not restricted to lockdowns. IoMT supports the ideas of *telehealth* and *telemedicine*. Keeping track of vital signs 24/7 with remote monitoring provides more details than a brief office visit and improves personalized diagnosis. More data and information enhance patient-doctor communication. Compared to manual alerts from traditional personal emergency response systems, IoMT devices can automatically alert medical personnel if something happens, e.g., a specific value falls below or exceeds a pre-defined threshold.

Despite this positive outlook, *Medical Device Manufacturers* (MDMs), *Healthcare Delivery Organizations* (HDOs), and patients must consider the inherent risks associated with this technology. Remote monitoring and control can increase patients' quality of life but also pose potential threats. According to Claroty [7], IoT vulnerability disclosures increased by 57% in the first half year of 2022. The FBI [12] warns HDOs about unpatched and outdated medical devices. IoMT can directly or indirectly influence patients' conditions. According to Ajagbe et al. [1], security is one of the major challenges of IoMT. It is difficult to monitor and keep devices up to date for a lifetime of up to ten years. If a component breaks down or multiple connections cause a deadlock, the clinical function of the device should still be operational.

In our previous work [31,33], we implemented context-aware security modes for medical devices and switched them based on vulnerability scores. For example, *Mode 0* provided core functionality and *Mode 1-3* extended functionality like remote monitoring and control. We have focused on securing single devices such as pacemakers or insulin pumps. Switching modes provides a method to reduce the attack surface. However, in light of the increasing number of IoMT, protecting and securing them requires a broader focus than that of a single device, as we have shown for IoT devices in [32]. As the devices are networked, this can also be used for security purposes. Some anomalies and attacks can be detected only or easier with multiple IoMT devices and a central control component.

In this paper, we propose the design of a *Self-protective MCPS*. We extend our previous work with a client-server perspective, multiple IoMT devices, and an *Intrusion Detection and Prevention System* (IDPS). Our work aims to resiliently protect patients, MCPSs, IoMT devices, and the environment by monitoring and automatically adapting when anomalies occur or vulnerabilities become known. Security and reactions to attacks is necessary on multiple layers. Depending on IoMTs' context, e.g., the connection state (online/offline), the reaction can be less or more restrictive.

The paper is structured as follows. In Sect. 2, we discuss related work. We describe the security and privacy challenges of MCPS and how to deal with them in Sect. 3. We present our proposal for a *Self-protective MCPS* architecture in Sect. 4. In Sect. 5, we show a sample scenario and discuss the implications in Sect. 6. Finally, we draw our conclusions in Sect. 7.

2   Related Work

In their vision of autonomic computing, Kephart & Chess [22] consider the concept of self-protection. In contrast to manually detecting and recovering from attacks by IT security professionals, self-protective systems can automatically defend against attacks, provide early warning, and attack mitigation methods. Likewise, self-healing capabilities [5] can enable systems to recover from attacks and reestablish functionalities. *Feedback* or *closed-loop systems* work similarly. Hellerstein et al. [19] consider feedback and sensor values to adapt systems to a goal without human intervention.

Our research builds up on the trustworthy multi-modal design for life-critical systems by Rao et al. [28]. They suggest decomposing systems into several modes facing different security risk values. Modes are switched based on events, system changes, or environmental changes related to risk values. While their focus during runtime is risk assessment for single devices, we consider sharing attacker information to prevent further attacks on other devices.

In the context of trustworthy secure systems, Ross et al. [34] suggest modes to encounter disruptions, hazards, and other threats. They describe modes for initialization, normal/operation/runtime, alternative, degraded, secure, standby, maintenance, training, simulation, test, recovery, shutdown/halted, and others. Each mode has its behavior, security configuration, and defined transitions to other modes. In addition, the German Federal Office for Information Security [4] differentiates among modes for medical operation, configuration, and technical maintenance in their cybersecurity requirements for network-connected medical devices.

3   Security and Privacy Challenges

Challenges in the medical domain have been addressed and discussed by several authors. The challenges include confidentiality, integrity, availability, reliability, safety, privacy, secure communication, software and hardware aspects, intrusion detection and reaction, formal methods, resource constraints, non-technical aspects, and organizational and regulatory issues [1, 9, 20, 21, 35, 37, 39].

The *NIST Cybersecurity Framework* [2] is a good starting point for these challenges. Based on existing standards, guidelines, and practices, it helps to manage and reduce cybersecurity risks with five core functions: *Identify, Protect, Detect, Respond, and Recover.* NIST also provides guidelines for foundational activities for IoT device manufacturers and a cybersecurity capability core baseline [10, 11].

In the USA, the *Food and Drug Administration* (FDA) provides reports, white papers for threat modeling, incident response, off-the-shelf software, patient communication, and guidance for pre- and postmarket cybersecurity and quality system considerations in medical devices [14, 17]. In the EU, the *Medical Device Coordination Group* provides guidance to fulfill the regulatory requirements [23]. The US Presidential Executive Order 14028, "Improving the Nation's Cybersecurity", pushed US agencies like the FDA to enhance cybersecurity and software supply chain security [38]. One result was the *Cybersecurity Modernization Action Plan* [18], which considers a zero trust approach, promotes best practices for secure development, and how to utilize *Artificial Intelligence/Machine Learning* (AI/ML) technologies for detection and response. In this paper, we focus on the challenges of *Reliability and Availability*, *Safety*, and *Malware and Intrusion Detection* and *Reaction*.

4 Self-protective MCPS Architecture

MCPS and connected IoMT devices should be designed to detect and respond to anomalies and potential cyberattacks and recover from them. Therefore, MDMs and HDOs need a more comprehensive view than just monitoring single devices. Considering events of multiple IoMT devices provides a better overview and reduces false positives. Additionally, the reaction to attacks can be implemented on multiple layers of the *Self-Protective MCPS*.

Figure 1 shows our proposed architecture. We follow the principle of divide and conquer and want to utilize a distributed MAPE-K loop, as presented in [32]. Decisions are made as de-centralized as possible and centralized as necessary. A *Manager* application will use its *Inventory* to continuously monitor connected *IoMT devices* to provide enhanced visibility and situational awareness for the *Operator*. The *Operator*, like an IT security professional, can analyze the situation and plan and execute adaptations to reach the system goals. For efficiency, repetitive tasks and decisions can be partially or fully automated. Additionally, the *Manager* centrally monitors *Common Vulnerabilities and Exposures* (CVEs), medical advisories, safety communications, product alerts, warnings, recalls, and other events of public databases, as we presented in our previous work [31,33]. Based on that, the *Manager* can automatically send adaptation requests to IoMT devices to change their behavior, e.g., block IP addresses or switch their mode. Likewise, the *Operator* and the *Patient* can do that manually.

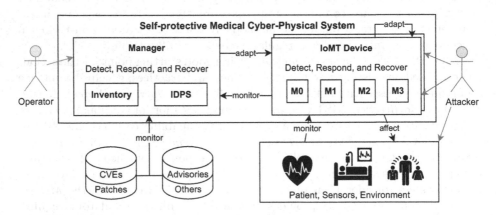

Fig. 1. Self-Protective Medical Cyber-Physical Systems Architecture.

Attackers can attack the *Self-Protective MCPS* on the hardware, software, and network layer. Therefore, anomaly and attack detection and reaction must be implemented on multiple levels. For example, trained deep autoencoder models on typical non-malicious network packets/flows can help to detect anomalous traffic [30]. *IoMT devices* monitor and affect their environment: the patient, attached sensors, and actuators. If IoMT devices are online, they can send data to the connected *Manager* and forward information and decisions about further actions. Depending on the configuration, the *Manager* automatically decides, or an *Operator* (human-in-the-loop) decides on further steps. If an IoMT device is offline, it has to make decisions on its own and adapt itself if necessary. Pre-defined rules and actions on the IoMT device help to achieve that. Additionally, software and hardware window watchdog timers may help to reset the system or components if the software crashes or hangs from a denial of service attack [36]. The ultimate goal is to provide clinical function while reducing the attack surface. To reach this goal, we leverage the multi-modal architecture by [31] and extend it with an IDPS, a lightweight version on the *IoMT devices*, and a more extensive version on the *Manager*.

For example, attackers who try to crack passwords, login credentials, and encryption keys can be blocked after multiple wrong attempts with local firewall rules. However, if these attacks reach a specified amount, affect availability, or lead to battery depletion, the IoMT device may adapt itself and switch to a more restrictive mode. We suggest a low-power mode with limited functions to extend battery life and reduce the attack surface. An activity sensor or timer can trigger switching to a mode with more functionality. During the connection of the IoMT device with the *Manager*, switching to the high-security mode may provide an encrypted channel and make the device more resistant. In case of an attack, devices switch to degraded or failure mode, and self-healing capabilities [5] can enable systems to recover from attacks.

5   Sample Scenario

Self-protective MCPSs can be used in hospitals and at home for patients with chronic diseases like diabetes. Typically, such systems for *Automated insulin delivery* (AID) work partly or fully automated [25]. They consist of wearable devices to monitor vital signs, continuous glucose monitors, wireless connected medicine pumps, and handheld devices or smartphones for local control and connection to the HDO. Based on device settings, patient history, and the current condition, the handheld analyzes the data and may adapt the settings. For example, if the blood glucose level changes, the system decides to increase or decrease the dose of medication. Within a specific threshold, this process is automated as a closed loop. A closed-loop system automatically considers feedback and sensor values to adapt to the system goal without human intervention [3]. There exist several commercial and non-commercial AID systems. In 2016, Medtronic MiniMed 670G was the first FDA-approved commercial AID system [13]. Before 2016, some affected patients did not want to wait any longer and developed *Do-It-Yourself* closed-loop implementations and provided them open source, but, needless to say, without warranty. According to Dana Lewis and the OpenAPS Community [8], over 2700 people still use this solution.

In our context of IoMT, the closed-loop scenario is extended with information transfer to the HDO for remote monitoring and reconfiguration by a physician, as described in Rao et al. [27]. Using asymmetric cryptography protocols for command and control messages, e.g., signed with a unique public key of each IoMT device, can secure communication [40]. If a measured value is outside pre-defined thresholds or if certain sequences of commands are unusual and potentially cause physical damage, cf. Stuxnet [6], or potentially harm patients, the *Operator* (human-in-the-loop) will get a notification. Then the data can be reviewed and affected settings adapted, e.g., switching from mode M1 to M0.

We had a closer look at the Medtronic MiniMed 600 series and its vulnerabilities and analyzed how a *Self-protective MCPS* would be beneficial. In 2022, Medtronic [24] alerted patients about a vulnerability in the protection mechanism in their MiniMed 600 series: Exploitation could compromise communication, allow unauthorized users to change the insulin delivery, and could "potentially lead to seizure, coma or death". According to the FDA recalls [15,16], over 600 000 products in commerce were affected. The company recommended that patients should manually turn off the "Remote Bolus" feature on the pump, which was on by default. Using our *Self-protective MCPS*, we would have simplified this step for patients. The insulin pump would have a connected and disconnected mode. In the disconnected mode, the pump works offline and considers only pre-defined presets. Additionally, manual changes using the switches on the physical hardware are possible. The disconnected mode is also the fallback mechanism if the connection to other devices gets lost. In the connected mode, the pump would consider the information of connected sensors and automatically adapt the medication dose. The *Manager* would have recognized the vulnerability, notified the *Operator*, and may suggest actions to adapt the IoMT devices, like installing patches or updates or switching from the connected to the disconnected mode. Using the inventory would allow the *Operator* to notify patients directly at the device and obtain consent before executing the interventions. If no update is available and the patient safety risk is too high, we would switch devices to the disconnected mode. Additionally, security-concerned patients could manually switch from the connected to the disconnected mode in general or as needed, for example, when they are away from home. In the successor product MiniMed 770G, Medtronic [26] included the auto and the manual mode to provide similar functionality.

Another recommendation of the MDM [24] was to connect or link devices only in private places. Switching to a connected and protected mode would be beneficial after the system's initial setup. Only pre-defined connections to trusted devices are allowed in this mode, but no new ones to reduce the attack surface. Additionally, the lightweight IDPS on the IoMT device could analyze the traffic and data from connected devices, notify the *Patient* and the *Operator* about abnormal behavior, and automatically delete the suspicious device from the trusted list. Sharing information about potential attacks like wrong connection attempts or abnormal behavior will enrich the security visibility for the *Operator*. For example, if an attacker tries to attack multiple IoMT devices, the *Manager* could recognize that, inform the *Operator*, and automatically warn other devices to increase the monitoring or to adapt security settings, e.g., by switching modes.

6 Discussion

Turning off the main features of healthcare systems is never an easy step and must only be considered as a last resort. *Self-protective MCPSs* can be a way to overcome this situation. Instead of just having the option to turn on and off devices, the availability of multiple modes provides more flexibility. A central *Manager* can allow HDOs to communicate with connected IoMT devices, analyze the security situation, notify patients (in specific cases), provide patches, and adapt IoMT device settings. Modes and mode switching, in turn, can pose new risks. We must take precautions so that malicious insiders cannot get control of the *Manager* and harm patients from this end. Thus, both technical and organizational security measures are essential.

The monitoring and control options are limited if the IoMT device has no connection to the *Manager*. A lightweight IDPS on IoMT devices can be beneficial by blocking suspicious traffic. However, in case of incorrect or faulty detection, this can lead to limited functionality. Another aspect results from the autonomy of *Self-protective MCPSs* itself. In highly automated scenarios, some serious events may remain undetected in the abundance of data and blind trust in the system.

7 Conclusion

Healthcare systems with connected IoMT devices pose many security threats and have to address several security and privacy challenges. We suggest taking advantage of their interconnected topology. Analyzing and correlating issues from multiple IoMT devices reveal anomalies and attacks that one device would not have recognized. In our *Self-protective MCPS* architecture, a central manager with an intrusion detection and prevention component can take over work from IoMT devices, analyze issues, and automatically take actions to adapt devices and prevent further attacks. Additionally, lightweight components on the IoMT devices can mitigate attacks if the device is offline. Our sample scenario has provided a first impression. We are now in the process of implementing our proposed architecture to experiment and simulate how it reacts to different attacks.

Acknowledgement. This work has partially been supported by the LIT Secure and Correct Systems Lab funded by the State of Upper Austria, the Austrian Marshall Plan Foundation, and the National Science Foundation under Grant Number 1622589 "Time-Centric Modeling of Correct Behaviors for Efficient Non-intrusive Runtime Detection of Unauthorized System Actions." Any opinions, findings, conclusions, or recommendations expressed in this material are those of the authors and do not necessarily reflect the views of the supporting organizations.

References

1. Ajagbe, S.A., Awotunde, J.B., Adesina, A.O., Achimugu, P., Kumar, T.A.: Internet of Medical Things (IoMT): applications, challenges, and prospects in a data-driven technology. In: Chakraborty, C., Khosravi, M.R. (eds.) Intelligent Healthcare. Springer, Singapore (2022). https://doi.org/10.1007/978-981-16-8150-9_14
2. Barrett, M.: Framework for Improving Critical Infrastructure Cybersecurity Version 1.1. No. NIST CSWP 04162018, U.S. National Institute of Standards and Technology (NIST), Gaithersburg, MD (2018). https://doi.org/10.6028/NIST.CSWP.04162018
3. Boughton, C.K., Hovorka, R.: New closed-loop insulin systems. Diabetologia **64**(5), 1007–1015 (2021). https://doi.org/10.1007/s00125-021-05391-w
4. BSI: Cyber Security Requirements for Network-Connected Medical Devices. German Federal Office for Information Security (BSI) (2018). https://www.bsi.bund.de/SharedDocs/Downloads/EN/BSI/ICS/Medical_Devices_CS-E_132.html. Accessed 28 Dec 2022
5. Carreon-Rascon, A.S., Rozenblit, J.W.: Towards requirements for self-healing as a means of mitigating cyber-intrusions in medical devices. In: 2022 IEEE International Conference on Systems, Man, and Cybernetics (SMC), pp. 1500–1505 (2022). https://doi.org/10.1109/SMC53654.2022.9945507
6. Chen, T.M., Abu-Nimeh, S.: Lessons from Stuxnet. Computer **44**(4), 91–93 (2011). https://doi.org/10.1109/MC.2011.115
7. Claroty: State of XIoT Security Report (2022). https://claroty.com/press-releases/iot-vulnerability-disclosures-grew-57-percent-from-2h21-to-1h22. Accessed 28 Dec 2022
8. Lewis, D., The OpenAPS Community: OpenAPS Outcomes (2022). https://openaps.org/outcomes/. Accessed 10 Jan 2023
9. Elhoseny, M., et al.: Security and privacy issues in medical internet of things: overview, countermeasures, challenges and future directions. Sustainability **13**(2121), 11645 (2021). https://doi.org/10.3390/su132111645
10. Fagan, M., Megas, K.N., Scarfone, K., Smith, M.: Foundational cybersecurity activities for IoT device manufacturers. No. NIST IR 8259, U.S. National Institute of Standards and Technology (NIST), Gaithersburg, MD (2020). https://doi.org/10.6028/NIST.IR.8259
11. Fagan, M., Megas, K.N., Scarfone, K., Smith, M.: IoT device cybersecurity capability core baseline. No. NIST IR 8259A, U.S. National Institute of Standards and Technology (NIST), Gaithersburg, MD (2020). https://doi.org/10.6028/NIST.IR.8259a
12. FBI: Industry Alert: Unpatched and Outdated Medical Devices Provide Cyber Attack Opportunities. U.S. Federal Bureau of Investigation (FBI) (2022). https://www.ic3.gov/Media/News/2022/220912.pdf. Accessed 28 Dec 2022
13. FDA: FDA approves first automated insulin delivery device for type 1 diabetes. U.S. Food and Drug Administration (FDA) (2016). https://www.fda.gov/news-events/press-announcements/fda-approves-first-automated-insulin-delivery-device-type-1-diabetes. Accessed 10 Jan 2023
14. FDA: Postmarket Management of Cybersecurity in Medical Devices. U.S. Food and Drug Administration (FDA) (2016). https://www.fda.gov/media/95862/download. Accessed 28 Dec 2022
15. FDA: Class 2 Device Recall Medtronic MiniMed 600 Series Insulin Pump Systems. U.S. Food and Drug Administration (FDA) (2022). https://www.accessdata.fda.gov/scripts/cdrh/cfdocs/cfRES/res.cfm?id=196205. Accessed 10 Jan 2023

16. FDA: Class 2 Device Recall Medtronic MiniMed 600 Series Insulin Pump Systems. U.S. Food and Drug Administration (FDA) (2022). https://www.accessdata.fda.gov/scripts/cdrh/cfdocs/cfRES/res.cfm?id=196183. Accessed 10 Jan 2023

17. FDA: Cybersecurity in Medical Devices: Quality System Considerations and Content of Premarket Submissions - Draft Guidance. U.S. Food and Drug Administration (FDA) (2022). https://www.fda.gov/media/119933/download. Accessed 28 Dec 2022

18. FDA: Cybersecurity Modernization Action Plan. U.S. Food and Drug Administration (FDA) (2022). https://www.fda.gov/media/163086/download. Accessed 28 Dec 2022

19. Hellerstein, J., Diao, Y., Parekh, S., Tilbury, D.: Feedback Control of Computing Systems. Wiley (2004). https://doi.org/10.1002/047166880X

20. IMDRF: Principles and Practices for Medical Device Cybersecurity. International Medical Device Regulators Forum (IMDRF) (2020). http://www.imdrf.org/docs/imdrf/final/technical/imdrf-tech-200318-pp-mdc-n60.pdf. Accessed 28 Dec 2022

21. Kagita, M.K., Thilakarathne, N., Gadekallu, T.R., Maddikunta, P.K.R.: A review on security and privacy of internet of medical things. In: Ghosh, U., Chakraborty, C., Garg, L., Srivastava, G. (eds.) Intelligent Internet of Things for Healthcare and Industry. Internet of Things. Springer, Cham (2022). https://doi.org/10.1007/978-3-030-81473-1_8

22. Kephart, J., Chess, D.: The vision of autonomic computing. Computer **36**(1), 41–50 (2003). https://doi.org/10.1109/MC.2003.1160055

23. MDCG: Guidance on Cybersecurity for medical devices. Medical Device Coordination Group (MDCG) (2019). https://ec.europa.eu/docsroom/documents/41863/attachments/1/translations/en/renditions/native. Accessed 28 Dec 2022

24. Medtronic: Urgent Medical Device Correction: MiniMed^{TM} 600 Series Pump System Communication Issue (2022). https://www.medtronicdiabetes.com/customer-support/product-and-service-updates/notice19-letter. Accessed 10 Jan 2023

25. Medtronic: MiniMed 670G System Discontinuation of New Sales (2023). https://www.medtronicdiabetes.com/products/minimed-670g-insulin-pump-system. Accessed 10 Jan 2023

26. Medtronic: The MiniMed 630G and 770G Insulin Pumps (2023). https://www.medtronic.com/us-en/healthcare-professionals/therapies-procedures/diabetes/education/diabetes-digest/minimed-insulin-pumps.html. Accessed 10 Jan 2023

27. Rao, A., Carreón, N.A., Lysecky, R., Rozenblit, J.: FIRE: a finely integrated risk evaluation methodology for life-critical embedded systems. Information **13**(1010), 487 (2022). https://doi.org/10.3390/info13100487

28. Rao, A., Rozenblit, J., Lysecky, R., Sametinger, J.: Trustworthy multi-modal framework for life-critical systems security. In: Proceedings of the Annual Simulation Symposium, ANSS 2018, San Diego, CA, USA, pp. 1–9. Society for Computer Simulation International (2018). https://doi.org/10.5555/3213032.3213049

29. Reports And Data: Market value of the internet of medical things worldwide in 2019 and 2027 (in billion U.S. dollars). Statista (2021). https://www.statista.com/statistics/1264333/global-iot-in-healthcare-market-size/. Accessed 28 Dec 2022

30. Rezvy, S., Petridis, M., Lasebae, A., Zebin, T.: Intrusion detection and classification with autoencoded deep neural network. In: Lanet, J.-L., Toma, C. (eds.) SECITC 2018. LNCS, vol. 11359, pp. 142–156. Springer, Cham (2019). https://doi.org/10.1007/978-3-030-12942-2_12

31. Riegler, M., Sametinger, J., Rozenblit, J.W.: Context-aware security modes for medical devices. In: 2022 Annual Modeling and Simulation Conference (ANNSIM), pp. 372–382 (2022). https://doi.org/10.23919/ANNSIM55834.2022.9859283

32. Riegler, M., Sametinger, J., Vierhauser, M.: A distributed MAPE-K framework for self-protective IoT devices. In: 2023 IEEE/ACM 18th International Symposium on Software Engineering for Adaptive and Self-Managing Systems (2023). https://doi.org/10.1109/SEAMS59076.2023.00034

33. Riegler, M., Sametinger, J., Vierhauser, M., Wimmer, M.: A model-based mode-switching framework based on security vulnerability scores. J. Syst. Softw. **200**, 111633 (2023). https://doi.org/10.1016/j.jss.2023.111633

34. Ross, R., McEvilley, M., Carrier Oren, J.: Systems Security Engineering: Considerations for a Multidisciplinary Approach in the Engineering of Trustworthy Secure Systems. No. NIST SP 800-160, U.S. National Institute of Standards and Technology (NIST) (2016). https://doi.org/10.6028/NIST.SP.800-160

35. Sametinger, J., Rozenblit, J., Lysecky, R., Ott, P.: Security challenges for medical devices. Commun. ACM **58**(4), 74–82 (2015). https://doi.org/10.1145/2667218

36. Stajano, F., Anderson, R.: The grenade timer: fortifying the watchdog timer against malicious mobile code. In: Proceedings of 7th International Workshop on Mobile Multimedia Communications, MoMuC 2000, Waseda, Tokyo, Japan (2000). https://www.cl.cam.ac.uk/~fms27/papers/2000-StajanoAnd-grenade.pdf. Accessed 28 Dec 2022

37. Sun, Y., Lo, F.P.W., Lo, B.: Security and privacy for the internet of medical things enabled healthcare systems: a survey. IEEE Access **7**, 183339–183355 (2019). https://doi.org/10.1109/ACCESS.2019.2960617

38. The White House: Executive Order 14028: Improving the Nation's Cybersecurity (2021). https://www.whitehouse.gov/briefing-room/presidential-actions/2021/05/12/executive-order-on-improving-the-nations-cybersecurity/. Accessed 27 Dec 2022

39. Thomasian, N.M., Adashi, E.Y.: Cybersecurity in the Internet of Medical Things. Health Policy Technol. **10**(3), 100549 (2021). https://doi.org/10.1016/j.hlpt.2021.100549

40. Zeadally, S., Das, A.K., Sklavos, N.: Cryptographic technologies and protocol standards for Internet of Things. IoT **14**, 100075 (2021). https://doi.org/10.1016/j.iot.2019.100075

An Approach for Safe and Secure Software Protection Supported by Symbolic Execution

Daniel Dorfmeister[1] , Flavio Ferrarotti[1]([✉]) , Bernhard Fischer[1] ,
Evelyn Haslinger[2], Rudolf Ramler[1] , and Markus Zimmermann[2]

[1] Software Competence Center Hagenberg, Hagenberg im Mühlkreis, Austria
{daniel.dorfmeister,flavio.ferrarotti,bernhard.fischer,
rudolf.ramler}@scch.at
[2] Symflower GmbH, Linz, Austria
{evelyn.haslinger,markus.zimmermann}@symflower.com

Abstract. We introduce a novel copy-protection method for industrial control software. With our method, a program executes correctly only on its target hardware and behaves differently on other machines. The hardware-software binding is based on Physically Unclonable Functions (PUFs). We use symbolic execution to guarantee the preservation of safety properties if the software is executed on a different machine, or if there is a problem with the PUF response. Moreover, we show that the protection method is also secure against reverse engineering.

1 Introduction

Industrial-scale reverse engineering is a serious problem, with estimated annual losses for industry at 6.4 billion euros in Germany alone[1]. Typically, the main effort needed to steal the intellectual property (IP) of companies producing machines with key software components resides in replicating the hardware. Contrary, software can often be copied verbatim with no reverse engineering required.

In the DEPS[2] (short for Dependable Production Environments with Software Security) project we investigate new approaches to prevent the described IP theft. In this paper, we explore binding a given program P to a specific target machine M, so that P only executes correctly on M. If P runs on a machine M'

---

[1] VDMA Product Piracy 2022 (https://www.vdma.org/documents/34570/51629660/ VDMA+Study+Product+Piracy+2022_final.pdf). Last accessed: 30/01/2023.
[2] https://deps.scch.at.

The research reported in this paper has been funded by BMK, BMDW, and the State of Upper Austria in the frame of the COMET Module Dependable Production Environments with Software Security (DEPS) within the COMET - Competence Centers for Excellent Technologies Programme managed by Austrian Research Promotion Agency FFG.

G. Kotsis et al. (Eds.): DEXA 2023 Workshops, CCIS 1872, pp. 67–78, 2023.
https://doi.org/10.1007/978-3-031-39689-2_7

other than M (even if M' is a replica of M), then P should behave differently, but still meet the required safety constraint for industrial applications. Our objective is to make the task of reverse engineering the protection extremely difficult and time consuming, rendering it uneconomical for the attacker.

We make use of Physically Unclonable Functions (PUFs), a hardware-based security primitive [6,8,14]. Minor variations in the manufacturing process of a hardware component cause unintended physical characteristics. Thus, this unique *digital fingerprint* cannot be cloned easily. A PUF uses these unique hardware characteristics to provide hardware-specific responses to user-defined challenges. PUFs can use designated hardware, but they can also be based on standard components like Dynamic Random Access Memory (DRAM) [9,10,18,19]. A common application of PUFs is for hardware-software binding [7,12,13,20].

In the environment targeted by DEPS, it is important that the alternative behavior of a protected program is always safe, in the sense that the program may run differently on different than the original hardware (making it difficult for an attacker to reverse engineer the protection), but it should do so without producing any harm. Moreover, if a protected program running on the target machine receives an (unlikely but nevertheless possible) incorrect response to a given PUF challenge, the program must still behave safely. Through the approach presented in this paper, we can attain this high degree of safety for a wide class of algorithms by applying symbolic execution techniques. At the same time, we can also ensure a high level of security, in the sense that reverse engineering the protected software becomes an extremely difficult (and expensive) task, without guarantee of success.

Xiong et al. [20] have previously presented a software protection mechanism in which they use dynamic PUFs (based on DRAM). They bind software to hardware, protecting it against tampering by considering the timing of the software for the PUF response. For the protection itself, they use self-checksumming code instances. With the help of a PUF response, the checksum, and a reference value, they determine a jump address. This can, however, result in unsafe behavior of the software or a crash, since the jump can be to a random function if the response of the PUF is not the expected one. We follow a different safe-by-design approach, where we only allow moves to next states that preserve safety constraints.

The verification of software safety by means of symbolic execution is not new—see, e.g., [1]. However, to the best of our knowledge, it has not been used to support safe copy-protection using software to hardware binding yet.

The paper is organized as follows: In Sect. 2, we briefly introduce the class of algorithms targeted by our protection method. This class is captured by control state Abstract State Machines (ASMs), which is also a convenient formal specification method to precisely describe the proposed protection mechanism. Note that ASMs can be considered as executable abstract programs and thus can also be symbolically executed. We specify our threat model in Sect. 3. Our main contribution is condensed in Sect. 4, where we introduce the proposed protection mechanism through an example and then generalize it to turn any ASM specification of a control state algorithm into a copy-protected (safe) specification bound to a target hardware through a suitable PUF. This is followed by a preliminary evaluation of the security of our protection method in Sect. 5. We conclude our presentation with a brief summary and future research plans in Sect. 6.

2 Control State ASMs

We use Abstract State Machines (ASMs) to formally specify the proposed copy-protection mechanism and to show that it is applicable to the whole class of algorithms captured by *control state* ASMs [5]. This is a particularly frequent class of ASMs that represents a normal form for UML activity diagrams. It allows the designer to define control Finite-State Machines (FSMs) with synchronous parallelism and the possibility of manipulating data structures. Moreover, indus-trial control programs (i.e., our target in DEPS) belong to this class [4]. This paper can be understood correctly by reading ASM rules as pseudocode over abstract data types. Next, we briefly review some of the basic ASM concepts. Standard reference books for ASMs are [4,5].

An ASM of some signature Σ can be defined as a finite set of transition rules of the form **if** *Condition* **then** *Updates*, which transforms states. The condition or guard under which a rule is applied is a first-order logic sentence of signature Σ. *Updates* is a finite set of assignments of the form $f(t_1,\ldots,t_n) := t_0$, which are executed in parallel. The execution of $f(t_1,\ldots,t_n) := t_0$ in a given state S proceeds as follows: (1) all parameters $t_0, t_1, \ldots, t_n$ are assigned their values, say $a_0, a_1, \ldots, a_n$, (2) the value of $f(a_1, \ldots, a_n)$ is updated to a_0, which represents the value of $f(a_1, \ldots, a_n)$ in the next state. Pairs of a function name f (fixed by the signature) and optional arguments $(a_1, \ldots, a_n)$ of dynamic parameter values a_i, are called locations. They are the ASM concept of memory units, which are abstract representations of memory addressing. Location value pairs (l, a) are called updates, the basic units of state change.

The notion of an ASM run or *computation* is an instance of the classical notion of the computation of transition systems. An ASM computation step in a given state simultaneously executes all updates of all transition rules whose guard is true in the state. If and only if these updates are consistent, the result of their execution yields a next state. A set of updates is consistent if it does not contain pairs $(l, a), (l, b)$ of updates to a common location l with $a \neq b$. Simultaneous execution, as obtained in one step through the execution of a set of updates, provides a useful instrument for high-level design to locally describe a global state change. Non-determinism, usually applied as a way of abstracting from details of scheduling of rule executions, can be expressed by the rule **choose** x **with** φ **do** r, which means that r should be executed with an arbitrary x chosen among those satisfying the property φ. We sometimes use the convenient rule **let** $x = t$ **in** r, meaning: assign the value of t to x and then execute r.

A *control state* ASM (see Def. 2.2.1. in [5]) is an ASM whose rules are all of the form defined in Listing 1.1. Note that if there is no condition $cond_i$ satisfied for a given control state i, then these machines do not switch state. There can only be a finite number of $ctlState \in \{1, \ldots, m\}$. In essence, they act as internal states of an FSM and can be used to describe different system *modes*.

```
1  if  ctlState = i  then
2      rule
3      if  cond₁  then
4          ctlState := j₁
5      ...
6      if  condₙ  then
7          ctlState := jₙ
```

Listing 1.1. General Form of Control State ASM Rules

Industrial control programs are usually required to satisfy certain safety and security properties. The main challenge for the copy protection method proposed in this paper is to ensure that these properties are maintained. Our method achieves this using symbolic execution [11,15]. Symbolic execution applies to ASMs the same way it does to high-level programming languages [16]. This is not surprising as ASMs are executable abstract programs [3].

Symbolic execution abstractly executes a program covering multiple possible inputs of the program that share a particular execution path through the code. In concrete terms, instead of considering normal runs in which all locations take actual values from the base set of the ASM states, we consider runs in which some state locations may take symbols (depicted by lowercase Greek letters) representing arbitrary values. ASM runs will proceed as usual, except that the interpretation of some expressions will result in symbolic formulas.

3 Threat Model

Our threat model focuses on IP theft. More precisely, on reverse engineering the obfuscated flow of a program. In this work, we assume a kind of white-box man-at-the-end scenario where an attacker is able to successfully gain access to an arbitrary number of protected programs. The attacker does not have direct access to the source code but can analyze and decompile the binaries to obtain it. They can also obtain the exact specification of the hardware on which the program runs and rebuild it. The attacker's goal is to obtain information about the correct program flow and to rebuild or copy it.

This threat model and the related security of the protection mechanism presented in the following will be further discussed in Sect. 5. We also address what happens when the attacker gains access to a protected system, i.e., to the hardware the software is bound to.

4 PUF-Based Software Protection Method

This section introduces the key ideas and contribution of the paper. In Subsect. 4.1, we present our novel approach to copy-protect programs specified as control state ASMs through a simple example. This illustrates our proposal to use PUFs to bind the correct execution of a control program to a target machine. Moreover, it illustrates how symbolic execution can be used to ensure that the protected control ASM satisfies all safety constraints of the original control ASM. In Subsect. 4.2, we condense the approach into a precise method that can be automated. Given an appropriate PUF, our method can turn any given control state ASM specification into an equivalent specification that will only run correctly on the target machine and satisfy the required safety constraints.

For our approach to work, the following requirements must be met by the chosen PUF:

- It must be possible to query the PUF at run-time and its response time should not adversely affect the functionality of the target program.
- No error correction is required, but the response of the PUF must be reasonably reliable so that the software executes correctly on the target machine most of the time. Ultimately, this will depend on the fault-tolerance requirement of the specific application.
- The number of different challenge-response pairs of the PUF should ideally be greater than or equal to the number of control states of the program that one would like to protect with our method. A smaller number would also work, but it would make the protection less secure against reverse engineering.

4.1   Approach

We introduce our approach through a simple example. Consider the ASM specification in [4] (cf. Sect. 2) of a one-way traffic light control algorithm. The proper behavior of this algorithm is defined by the ASM rule in Listing 1.2. For this example, there are 4 possible control states as shown in Fig. 1. Only the first three are used by the correct specification. The fourth possible control state (i.e., $Go1Go2$) represents an unsafe, undesirable behavior.

```
1  1WayStopGoLight =
2  if  phase ∈ {Stop1Stop2, Go1Stop2} and Passed(phase) then
3        StopLight(1) := ¬StopLight(1)
4        GoLight(1) := ¬GoLight(1)
5        if  phase = Stop1Stop2 then
6              phase := Go1Stop2
7        else
8              phase := Stop2Stop1
9  if  phase ∈ {Stop2Stop1, Go2Stop1} and Passed(phase) then
10       StopLight(2) := ¬StopLight(2)
11       GoLight(2) := ¬GoLight(2)
12       if  phase = Stop2Stop1 then
13             phase := Go2Stop1
14       else
15             phase := Stop1Stop2
```

Listing 1.2. One-Way Traffic Light: Correct Specification

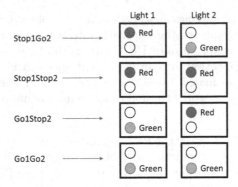

Fig. 1. Possible states of one-way stop-go traffic lights.

We propose to bind this control state algorithm to the hardware using the sub-rule CHOOSEPHASE in Listing 1.3, which updates the value of *phase* to the correct next phase shown in Listing 1.2 only if the algorithm is executed on the target hardware. Otherwise, CHOOSEPHASE updates *phase* by choosing non-deterministically among the set of *safePhases*, i.e., the phases that will not lead to *GoLight(1)* = *GoLight(2)* = *true*, and thus violate the safety constraint. This procedure ensures that if the software is copied verbatim to a machine other than the target one, it will behave safely but also incorrectly and nondeterministically, making it challenging for an attacker to reverse engineer the protection.

```
1  CHOOSEPHASE =
2  let  nextPhase = queryPUF(phase)  in
3      if  nextPhase ∈ safePhases  then
4          phase := nextPhase
5      else
6          choose  x ∈ safePhases  do
7              phase := x
```

Listing 1.3. Software-to-Hardware Binding Procedure

We use symbolic execution to systematically determine the set of *safePhases*, specifically (1) to determine the necessary conditions for a next state to be safe, and (2) to verify that this is indeed the case. Note that we must ensure that the protection mechanism will not introduce any bug that leads to an unsafe state. First, we assign a symbolic value to all relevant state locations:

$$L_0 = \{phase \mapsto \alpha, Passed(phase) \mapsto \beta, StopLight(1) \mapsto \gamma, GoLight(1) \mapsto \neg\gamma,$$
$$StopLight(2) \mapsto \delta, GoLight(2) \mapsto \neg\delta\}$$

Note that the relevant locations (or equivalently locations of interest) are all those that correspond to dynamic function names that appear in the conditional statements of 1WAYSTOPGOLIGHT. This might be in abundance since there could be locations that are not relevant to preserve the required safety constraint, i.e., that $\neg(GoLight(1) \wedge GoLight(2))$. The problem is that if we only consider a sub-set of locations, there is the possibility that we miss some execution path that leads to a violation of the safety constraint that we want to preserve. In general, this is undecidable and compromises might be necessary for complex applications.

Next, we look at all value path conditions resulting from executing the *protected* rule 1WAYSTOPGOLIGHT (i.e., PROTECTED1WAYSTOPGOLIGHT) obtained by replacing lines $6, 8, 13$ and 15 in 1WAYSTOPGOLIGHT with the sub-rule CHOOSEPHASE. W.l.o.g., in the symbolic execution we can simply assume that CHOOSEPHASE assigns a symbolic value α' (possibly different from α) to *phase*, since, at this point, we are only interested in determining the path conditions where the current value α of *phase* can lead to $GoLight(1) = GoLight(2) = true$ in the next state. This gives us 4 path conditions and 4 corresponding symbolic state locations, one for each of the 4 possible symbolic step transitions $a, b, c,$ and d (executions) of PROTECTED1WAYSTOPGOLIGHT.

- Path conditions P_a and set of symbolic state locations L_a of transition a:

$$P_a \equiv (\alpha = Stop1Stop2 \vee \alpha = Go1Stop2) \wedge \beta \wedge \alpha = Stop1Stop2$$
$$L_a = \{phase \mapsto \alpha', Passed(phase) \mapsto \beta, StopLight(1) \mapsto \neg\gamma, GoLight(1) \mapsto \gamma,$$
$$StopLight(2) \mapsto \delta, GoLight(2) \mapsto \neg\delta\}$$

- Path conditions P_b and set of symbolic state locations L_b of transition b:

$$P_b \equiv (\alpha = Stop1Stop2 \vee \alpha = Go1Stop2) \wedge \beta \wedge \alpha \neq Stop1Stop2$$
$$L_b = \{phase \mapsto \alpha', Passed(phase) \mapsto \beta, StopLight(1) \mapsto \neg\gamma, GoLight(1) \mapsto \gamma,$$
$$StopLight(2) \mapsto \delta, GoLight(2) \mapsto \neg\delta\}$$

- Path conditions P_c and set of symbolic state locations L_c of transition c:

$$P_c \equiv (\alpha = Stop2Stop1 \vee \alpha = Go2Stop1) \wedge \beta \wedge \alpha = Stop2Stop1$$
$$L_c = \{phase \mapsto \alpha', Passed(phase) \mapsto \beta, StopLight(1) \mapsto \gamma, GoLight(1) \mapsto \neg\gamma,$$
$$StopLight(2) \mapsto \neg\delta, GoLight(2) \mapsto \delta\}$$

- Path conditions P_d and set of symbolic state locations L_d of transition c:

$$P_d \equiv (\alpha = Stop2Stop1 \vee \alpha = Go2Stop1) \wedge \beta \wedge \alpha \neq Stop2Stop1$$
$$L_d = \{phase \mapsto \alpha', Passed(phase) \mapsto \beta, StopLight(1) \mapsto \gamma, GoLight(1) \mapsto \neg\gamma,$$
$$StopLight(2) \mapsto \neg\delta, GoLight(2) \mapsto \delta\}$$

In the next step, we group the path conditions that lead to the same symbolic state. In our running example, we see that whenever the condition $P_a \vee P_b$ is satisfied, we get the symbolic state $L_a = L_b$. Likewise, whenever the condition $P_c \vee P_d$ is satisfied, we get the symbolic state $L_c = L_d$. Using standard logical equivalences, we can simplify the path conditions as follows:

$$P_a \vee P_b \equiv (\alpha = Stop1Stop2 \vee \alpha = Go1Stop2) \wedge \beta$$
$$P_c \vee P_d \equiv (\alpha = Stop2Stop1 \vee \alpha = Go2Stop1) \wedge \beta$$

Moreover, if we restrict our attention to the locations of interest and their symbolic values following a step transition, we can clearly see that in the next state the assertion $GoLight(1) = GoLight(2) = true$ holds, i.e., the safety constraint is violated, iff $Passed(phase)$ and

$$((phase = Stop1Stop2 \lor phase = Go1Stop2) \land \neg GoLight(1) \land GoLight(2)) \lor$$
$$((phase = Stop2Stop1 \lor phase = Go2Stop1) \land \neg GoLight(2) \land GoLight(1))$$

Finally, we get that $safePhases$ is the set

$$\{x \in \{Stop1Stop2, Go1Stop2, Stop2Stop1, Go2Stop1\} \mid \neg\text{Cond}(x)\}$$

where

$$\text{Cond}(x) \equiv ((x = Stop1Stop2 \lor x = Go1Stop2) \land \neg GoLight(1) \land GoLight(2)) \lor$$
$$((x = Stop2Stop1 \lor x = Go2Stop1) \land \neg GoLight(2) \land GoLight(1)))$$

4.2   Method

We now generalize the example from the previous subsection to a precise method that can be automated. Let P be a control state ASM (i.e., an executable abstract program) with finitely many control states $ctlState \in \{1, \ldots, m\}$, whose rules $r_1, \ldots, r_m$ are all of the form defined in Listing 1.1. We assume w.l.o.g. that each r_i is guarded by **if** $ctlState = i$ **then** r_i. We define:

$$A = \{(i, j) \mid i, j \in \{1, \ldots, m\} \land subRuleOf(ctlState := j, r_i)\}$$

where $subRuleOf(s, r)$ is a Boolean function that evaluates to true only if s is a sub-rule of r. That is, $(i, j) \in A$ iff the rule r_i updates the control state to j.

The mechanical steps to copy-protect P with safety property Ψ using a (suitable) PUF $puf : C \to R$, with the set of challenges C and corresponding responses R, are described in the following 7 steps. Note that Step 1 is simply a prerequisite to implement the function $queryPUF$ used in Listing 1.3 in the one-way traffic lights example introduced in the previous section. Indeed, the CHOOSEPHASE rule results from applying Step 2 of our method. The next two steps in our method simply refine the application of the proposed protection based on updating control states via the PUF function, i.e., it corresponds to using CHOOSEPHASE to select the next control state in our previous example. Steps 5–7 are a formalization and generalization of the symbolic execution analysis done in the previous section to determine the safe control state transitions, that is to determine the set of $safePhases$ in our example.

1. Fix an injection $f : C' \to R$ such that $C' \subseteq C$, $|C'| = |A|$ and $f(x) = puf(x)$, and a corresponding bijection $g : A \to C'$. Note that this function f is simply an injective restriction of puf to a subdomain C' of size $|A|$, as required to encode all possible control state transitions of P (using the function g). If it is not possible to fix such an injective function f, this means that the

function *puf* is not "big enough" to encode all control state transitions. In this latter case, we simply consider that *puf* is not suitable for our protection strategy. An alternative would be to apply the protection to a subset of control states, but this would unnecessarily complicate the presentation of the general method.

2. For each tuple $(x, y) \in A$, define CHOOSECTLSTATE(x, y) as in Listing 1.4.

```
1  CHOOSECTLSTATE(x, y) =
2  let  nextCtlState = puf(g(x, y))  in
3        if  nextCtlState ∈ safeCtlStates  then
4              ctlState := nextCtlState
5        else
6              choose  z ∈ safeCtrlStates  do
7                    ctlState := z
```

Listing 1.4. Choose next control state using the PUF

3. For each control state rule r_i $(i = 1, \ldots, m)$ of P, replace every occurrence of a sub-rule of the form *ctlState* $:= j$ by the CHOOSECTLSTATE(*ctrlState*, j) sub-rule defined in the previous step.
4. In each control state rule r_i $(i = 1, \ldots, m)$ of P, replace the condition *ctlState* $= i$ in its guard by *ctlState* $\in \{x \mid \exists(y, z) \in A \, (g(y, z) = x \wedge y = i)\}$.
5. Assign a different symbolic value to all locations in the initial state that correspond to terms that appear in some conditional statements of a **if**-rule in P, i.e., to all locations of interest. This results in an assignment *symVal*, such that *symVal*$(h(\bar{t})) = \alpha_i$ iff α_i is the symbolic value assigned to the location $(h, \bar{t})$.
6. Symbolically execute P on the (symbolic) initial state built in the previous step. This results in a set *Sym* of pairs of the form (φ_i, L_i), where φ_i is a path condition (as in standard symbolic execution) and L_i is a set that maps each term (corresponding to a location with a symbolic value) to its symbolic value in the successor state of the initial one.
7. Finally, set *safeCtlStates* $= \{x \in \{1, \ldots, m\} \mid \neg \text{Cond}(x)\}$, where

$$\text{Cond}(x) \equiv \bigvee_{(\varphi_i, L_i) \in Sym} \left(\varphi_i' \wedge \Psi(L_i) \right)$$

and φ_i' is the expression obtained by replacing the symbolic values in φ_i by the terms corresponding to the locations with these symbolic values in the initial state. In turn, $\Psi(L_i)$ is obtained by first replacing in Ψ (i.e., in the safety constraint) every term t_j by the symbolic expression $L_i(t_j)$ and then replacing the symbolic values in the resulting expression by terms corresponding to the locations with these symbolic values in the initial state.

5   Security Evaluation

The safety of the proposed method for copy-protection of control state software is ensured by construction. We now analyze its security. That is, assuming the threat model in Sect. 3, we evaluate how an attacker could circumvent the copy-protection. To eliminate the protection, an attacker needs to perform an analysis that enables them to determine the correct value of $nextCtlState$ every time the rule in Listing 1.4 is called, or equivalently, whenever the current $ctlState$ is updated. The attacker can only access a binary of the target program. Thus, they will need to perform some kind of analysis, either of the binary, or of some decompiled version of it. The analysis can either be static or dynamic. Next, we discuss both possibilities.

Static Analysis. We assume that an attacker can gain access to any number of protected binaries, compiled using different PUFs, i.e., for different systems. This clearly does not provide any additional information regarding the values that $nextCtlState$ should take at different execution stages. Indeed, the attacker cannot determine the value of the function puf (see line 2 of Listing 1.4) for a given challenge with static analysis. Even if an attacker can decompile the binary, which is not a trivial task for machine code, and then bypasses any obfuscation applied to it and gets a human readable version of the protected program, this still will not provide any information regarding the challenge/response values of puf. An alternative is for the attacker to understand the logic of the control state program through inspection of a decompiled high-level code. Then they could possibly determine the correct $nextCtlState$ for each of the control rules. This task is time consuming and complex, even for the simple one-way traffic light control algorithm in Subsect. 4.1. In fact, writing the algorithm from scratch would probably be faster. This could easily become insurmountable as the complexity of the control state program increases and consequently its logic and number of states. We can conclude that pure static analysis is not a real threat by itself for our protection method.

Dynamic Analysis. We consider two scenarios for dynamic analysis. In the first one, the attacker has access to a system with identical specifications as the system the protected software is bound to. This does not give any advantage to the attacker over the static analysis, since the PUF cannot be cloned. Moreover, the runs of the program on this hardware will be non-deterministic w.r.t. $nextCtlState$.

In the second scenario, the attacker can run the program on the intended system, i.e., the unique system where the function puf will return the correct responses. Then, the attacker can trace the correct program flow via dynamic analysis. This can still be time consuming, in particular if the system is complex with many possible execution paths. Moreover, this dynamic analysis can interfere with PUF responses. For example, operations on DRAM are time-dependent, including the Rowhammer PUF [2,17], for instance, where memory must be accessed at high frequency, which would be hindered by other processes. The attacker could bypass this by querying the PUF separately from the protected program and then integrating the responses into the binary, but this would, once again, require a complex, time-consuming, and detailed analysis. This relative weakness can be dealt with by using complementary obfuscation techniques and/or physical protection mechanisms.

6 Summary

We introduced a novel method to bind control state software to specific hardware. The method ties the logic of the control state program to the unique responses provided by the PUF of the target hardware, so that it will only behave correctly if it is executed on the correct machine. Otherwise, the program will behave differently and in a non-deterministic manner. At the same time it will not crash, turning reverse engineering of the protection into a difficult, complex, and expensive task. Moreover, our copy-protection method ensures by construction that the safety properties of the software are preserved, even when it is illegally copied and executed on a cloned machine. This high level of safety is enabled by applying symbolic execution techniques and for a wide class of algorithms, namely for any algorithm that can be correctly specified by a control state ASM.

In future work, we plan to apply this protection method to an industrial case study, developing the necessary tools to automate the required tasks. This will allow us to evaluate aspects such as scalability of the proposed method and concrete effort required to reverse engineer it. It should be noted that step 7 of the method might result on long formulae whose evaluation could unacceptably affect the response time of a complex control state algorithm. In principle, this can be dealt with by simplifying the resulting expressions using logical equivalences (as in the example presented in Subsect. 4.1). Theorem provers in general, and SMT solvers in particular, can assist in this task. The latter have a long history of successful application in conjunction with symbolic execution.

In its current initial form, our method for software protection requires to manually identify the control states of the target programs as well as a formal specification of the safety constraints that need to be preserved. Processes to automate the identification of control states that are suitable to apply the described protection are currently being investigated within the DEPS project. Safety constraints necessarily need to be translated into a symbolic form so that symbolic execution can be applied to prove their preservation.

References

1. Ahmed, M., Safar, M.: Symbolic execution based verification of compliance with the ISO 26262 functional safety standard. In: DTIS 2019 (2019)
2. Anagnostopoulos, N.A., et al.: Intrinsic run-time Row Hammer PUFs: leveraging the row hammer effect for run-time cryptography and improved security. Cryptography **2**(3), 13 (2018)
3. Börger, E.: The role of executable abstract programs in software development and documentation. CoRR arXiv:2209.06546 (2022)
4. Börger, E., Raschke, A.: Modeling Companion for Software Practitioners. Springer, Heidelberg (2018). https://doi.org/10.1007/978-3-662-56641-1
5. Börger, E., Stärk, R.: Abstract State Machines. Springer, Heidelberg (2003). https://doi.org/10.1007/978-3-642-18216-7
6. Gassend, B. et al.: Silicon physical random functions. In: CCS 2002 (2002)

7. Guajardo, J., Kumar, S.S., Schrijen, G.-J., Tuyls, P.: FPGA intrinsic PUFs and their use for IP protection. In: Paillier, P., Verbauwhede, I. (eds.) CHES 2007. LNCS, vol. 4727, pp. 63–80. Springer, Heidelberg (2007). https://doi.org/10.1007/978-3-540-74735-2_5

8. Herder, C., et al.: Physical unclonable functions and applications: a tutorial. Proc. IEEE **102**(8), 1126–1141 (2014)

9. Keller, C., et al.: Dynamic memory-based physically unclonable function for the generation of unique identifiers and true random numbers. In: ISCAS 2014 (2014)

10. Kim, J.S., et al.: The DRAM latency PUF. In: HPCA 2018. IEEE (2018)

11. King, J.C.: Symbolic execution and program testing. Commun. ACM **19**(7), 385–394 (1976)

12. Kohnhäuser, F., Schaller, A., Katzenbeisser, S.: PUF-based software protection for low-end embedded devices. In: Conti, M., Schunter, M., Askoxylakis, I. (eds.) Trust 2015. LNCS, vol. 9229, pp. 3–21. Springer, Cham (2015). https://doi.org/10.1007/978-3-319-22846-4_1

13. Kumar, S.S. et al.: The butterfly PUF protecting IP on every FPGA. In: HOST 2008. IEEE (2008)

14. McGrath, T. et al.: A PUF taxonomy. Appl. Phys. Rev. **6**(1), 011303 (2019)

15. Pasareanu, C.S.: Symbolic Execution and Quantitative Reasoning: Applications to Software Safety and Security. Morgan & Claypool Publishers (2020)

16. Paun, V.A., Monsuez, B., Baufreton, P.: Integration of symbolic execution into a formal abstract state machines based language. IFAC-PapersOnLine **50**(1), 11251–11256 (2017)

17. Schaller, A., et al.: Intrinsic Rowhammer PUFs: leveraging the Rowhammer effect for improved security. In: HOST 2017. IEEE (2017)

18. Sutar, S., Raha, A., Raghunathan, V.: D-PUF: an intrinsically reconfigurable DRAM PUF for device authentication in embedded systems. In: CASES 2016 (2016)

19. Xiong, W., et al.: Run-time accessible DRAM PUFs in commodity devices. In: Gierlichs, B., Poschmann, A.Y. (eds.) CHES 2016. LNCS, vol. 9813, pp. 432–453. Springer, Heidelberg (2016). https://doi.org/10.1007/978-3-662-53140-2_21

20. Xiong, W. et al.: Software protection using dynamic PUFs. IEEE Trans. Inf. Forensics Secur. **15**, 2053–2068 (2019)

Towards Increasing Safety in Collaborative CPS Environments

Marco Stadler[1]([⊠])[iD], Michael Riegler[1,2][iD], and Johannes Sametinger[1,2][iD]

[1] LIT Secure and Correct Systems Lab, Johannes Kepler University, Linz, Austria
{marco.stadler,michael.riegler,johannes.sametinger}@jku.at
[2] Institute of Business Informatics – Software Engineering, Johannes Kepler University, Linz, Austria
https://www.jku.at/en/lit-secure-and-correct-systems-lab,
https://www.se.jku.at

Abstract. Cyber-Physical Systems (CPS) frequently operate in collaborative environments with other CPS and humans. This collaborative environment has the potential for situations in which CPS endanger humans. We argue that safety in such environments can be increased if the environment is aware of the safety-critical situation and can respond appropriately. In this paper, we describe our preliminary work on a collaborative CPS safety framework that combines distinct modes of operation with adaptive monitoring.

Keywords: Collaborative CPS · Adaptive Monitoring · Modes

1 Introduction

Cyber-Physical Systems (CPS) frequently operate in safety-critical environments and domains due to their close interaction with humans [4]. While recent efforts in securing the collaboration seem promising [12], safety incidents jeopardizing human well-being still occur. Reasons for these safety incidents range from security breaches [7] to malfunctioning systems due to system design flaws and sensor failures [6].

Alongside other countermeasures, the concept of *Modes* has been introduced to address this issue. Modes provide a set of functionalities to ensure a particular system behavior. We can switch modes based on certain circumstances. The trigger for switching between the modes depends on the context. From a functional standpoint, self-driving vehicles [3] use different modes to operate autonomously or manually. These triggers can also be based on safety risks; for instance, in the area of robotics, a manufacturing robot can switch between modes and adjust its movement speed based on the proximity of a human to avoid the risk of collision [18].

The detection of safety risks is not trivial. *Monitoring* certain properties of the CPS itself or the environment surrounding the CPS to detect potential safety risks as they occur is a crucial technique to accomplish this objective [9]. This

G. Kotsis et al. (Eds.): DEXA 2023 Workshops, CCIS 1872, pp. 79–85, 2023.
https://doi.org/10.1007/978-3-031-39689-2_8

is, for instance, the distance value measured by a LiDAR unit or a temperature measurement that prevents overheating in the preceding robotic example.

Furthermore, multiple CPS often operate as part of a *collaborative CPS*, to complete a specific mission, making the process of ensuring safety even more complex [1]. Recent efforts, therefore, have expanded this mode concept by facilitating the sharing of mode-related data between multiple CPS [16].

Most of the related approaches consider modes either for single systems [18], target only certain aspects like multi-mode real-time monitoring [11], or focus on formal frameworks for the design of safety monitors for multi-functional robotic systems with modes [5]. However, none of these approaches completely consider the combination of mode switches and adaptive monitoring for enhancing safety in a collaborative CPS environment.

The concept has thus far been employed in both security and safety contexts. In this paper, we present our initial efforts to increase safety in a collaborative CPS environment by combining the sharing of mode-related data with adaptive monitoring, thereby concentrating on the safety aspect of modes.

In detail, we claim the following contributions: (*i*) We present a list of challenges (*c.f.* Sect. 3) associated with environmental safety risks caused by multiple CPS collaborating with humans and (*ii*) derive an initial framework architecture (*c.f.* Sect. 4) utilizing mode switching and adaptive monitoring for mitigating these risks. In addition, we (*iii*) provide a *Proof of Concept* (PoC) of the framework (*c.f.* Sect. 5) to demonstrate the viability of our approach.

2   Motivation

CPS and robotic systems frequently operate in hazardous environments. For instance, accidents involving jamming, cutting, and crushing continue to occur frequently in industrial settings where collaborative work between multiple CPS and humans is prevalent, making these environments hazardous. Studies [8] indicate that the majority of incidents occurred during non-routine work, such as inspections, cleanings, or repairs, i.e., when systems are not operating in their typical mode(s) of operation. Therefore, systems undergoing non-routine modes of operations, such as *Maintenance Mode*, represent a safety-sensitive time frame and can be considered *safety-critical modes*. Due to its complexity, uncertainty, and variability, the environment of a CPS poses unique safety risks [1]. These risks are exacerbated when the collaborative CPS are operating additionally in a safety-critical mode. Fatal incidents are caused by the environment of collaborative CPS operating in a safety-critical mode. For instance, during an incident on a manufacturing floor [2], a human conducting maintenance tasks was killed by a robotic CPS from the environment that entered the maintenance zone by mistake. However, the incident could have been avoided if the environment had responded appropriately to the ongoing maintenance by switching the CPS in the environment into respective restrictive modes (e.g., completely disabled certain unsafe movements) and intensifying movement monitoring around the maintenance zone to detect collisions/malfunctions early on (i.e., employed adaptive monitoring). Therefore, we contend that the safety of such a collaborative CPS environment can be enhanced by leveraging appropriate mode switching and adaptive monitoring.

3 Challenges

Given a scenario in which multiple robotic manipulators are working in close proximity, one enters a *Maintenance Mode* as a worker proceeds to perform certain tasks on/near one of the manipulators. This condition comes along with a series of risks. The safety zones of the *System under Maintenance* (SuM) itself are violated, making a human operate within the operating zone of the robot and therefore susceptible to collisions. Monitoring properties that adhere to the detection of a potential collision has the utmost priority at this time. Therefore, the framework must be able to **adapt monitoring of the SuM alongside the mode switch (C1)** to ensure the prompt detection of potential collisions. Humans often disregard safety rules and systems malfunction. As a result, they may operate outside the safe range, endangering other systems (in this case, other manipulators or passing autonomous vehicles) or even themselves. The **environment must therefore transition to precautionary modes (C2)** that disable unsafe CPS behavior. The environment and the CPS themselves evolve (e.g., the manipulators receive new sensors, a new CPS is added to the factory floor, or a CPS is capable of driving into the safety-critical zone). Therefore, the framework must be **capable of adjusting to changes and co-evolve with the monitored systems (C3)**. The detection of environment-wide patterns is essential for environmental safety. Consequently, **data on changes in modes and monitoring of a CPS must be aggregated (C4)** to derive additional insights on the monitored environment, as certain patterns can only be detected at a higher level of abstraction (e.g., a system-wide failure due to power outages). Since safety incidents continue to occur, it is essential to **preserve data for post-mortem examination (C5)**. Based on the persisted data, incident scenarios must be revisited to derive alterations to the mode switching logic configuration and adaptive monitoring.

4 Framework Architecture

To address the aforementioned challenges, we present our preliminary work on a framework capable of adaptively monitoring CPS and switching modes in a collaborative CPS environment. An overview of the framework can be found in Fig. 1. The framework architecture was conceived based on the identified challenges (**C1–C5**). The correspondence between a specific challenge and a framework component is indicated with the blue ellipses.

The framework consists of five main components: `Environment`, `Registry`, `Communication Broker`, `Adaption Controller`, and `Services`. The `Environment` consists of all the CPS that might influence each other's safety. The

`Communication Broker` is intended to use a topic-based protocol providing a standardized interface capable of handling diverse systems. The topic-based architecture enables CPS to dynamically (un-) subscribe to changes in the `Environment`, therefore enabling a co-evolving framework (*c.f.* **C3**). The topics are assigned by a `Registry` that keeps track of where CPS are (physically) located (*c.f.* `Zone Registry`) and which CPS poses certain features (*c.f.* `CPS Registry`) that might influence the safety of an environment (e.g., a property indicating that a vehicle is capable of moving freely in the factory floor). Once a CPS enters a safety-critical mode (*c.f.* `CPS#2`) the information regarding the mode switch and meta-data corresponding to the switch are published via the `Communication Broker`. Based on the previously assigned topics, the respective CPS in the environment are notified via topic subscriptions (e.g., `CPS#1` and `CPS#3` are notified as they are nearby and `CPS#4` is notified as it might move into the safety-critical zone). CPS that are irrelevant (*c.f.* `CPS#5` and `CPS#6`) are neglected and operations continue as usual. The influential CPS in the environment and the CPS in the safety-critical mode then request an adaptation from the `Adaptation Controller`. Mode switches (*c.f.* **C2**) and a new monitoring configuration (*c.f.* **C1**) are provided by the `Mode Manager` and the `Monitoring Manager`. This information is consolidated by the `Core` and forwarded back to the respective CPS that can adapt accordingly. In parallel to this process, data collected by the `Environment` is aggregated at the `Aggregator` and the insights are fed into the `Adaptation Controller` to react accordingly (*c.f.* **C4**). Finally, all the gathered data concerning the mode switches and the CPS sensor data is persisted at the `Persistor` (*c.f.* **C5**).

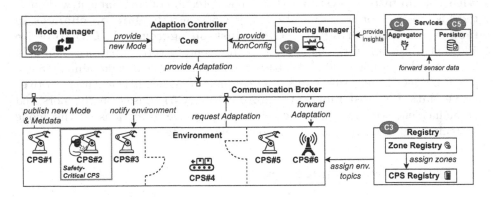

Fig. 1. High-level overview of the architecture.

5   Proof of Concept

To demonstrate the viability of the approach, we incorporated the proposed framework into a prototype PoC containing the core functionalities (i.e., the outlined core components, communication flows, mode switching logic, and adaptive monitoring). We used CPS employing the Robot Operating System (ROS) for the Python-based prototype and simulated a safety-critical mode transition in one of the CPS using TurtleBots [14] in a test scenario. The CPS in the environment then switched modes and adapted their monitoring behavior to reduce the time required to detect collisions in safety-critical zones. The source code is incorporated within a ready-to-use ROS package and available on GitHub[1].

6   Related Work

Collaborative CPS in safety-critical environments has previously been investigated. Zacharaki et al. [20], for instance, provide a systematic overview and characterization of safety features in human-robot interaction. They conclude that the runtime phase requires "novel, robust, and generalizable safety methods" to ensure the safe incorporation of these systems. This work is intended to contribute to this objective.

The two main concepts used in our approach (modes and adaptive monitoring) are employed in different CPS contexts. Yin and Hansson [19] address CPS complexity by leveraging a multi-mode system. Niu et al. [13] use modes to describe the system states of CPS undergoing malicious cyber attacks. Vierhauser et al. [17] provide a domain-specific language and framework for adaptive monitoring of CPS and Poltavtseva et al. [15] provide an adaptive information security monitoring system for CPS. Most of these approaches only use one concept, modes, or adaptive monitoring, while we argue in this paper that the combination of these conceptions yields great potential.

Malm et al. [10] present a dynamic safety system for industrial robots collaborating with humans but do not consider the environment of the collaborative systems.

Neukirchner et al. [11] use multi-mode monitoring for mixed-criticality real-time systems. While their work focuses predominantly on the provision of efficient real-time monitoring, our approach focuses on providing safety at a higher level of abstraction by considering the interaction of multiple CPS.

7   Conclusion

In this paper, we describe our initial efforts to increase safety in a collaborative CPS environment by employing mode switching and adaptive monitoring. We develop a general framework based on the challenges of such a safety-critical environment. A PoC is utilized to validate the viability of the proposed approach. Future efforts concentrate on the complete implementation of the framework and a case-study evaluation.

Acknowledgement. This work has been supported by the LIT Secure and Correct Systems Lab funded by the State of Upper Austria and the Linz Institute of Technology (LIT-2019-7-INC-316).

---

[1] https://github.com/jku-lit-scsl/mode-mon.

References

1. Ali, N., Hussain, M., Hong, J.E.: Analyzing safety of collaborative cyber-physical systems considering variability. IEEE Access **8**, 162701–162713 (2020)
2. Baldas, T.: Lawsuit: Defective robot killed factory worker; human error to blame (2017). https://eu.freep.com/story/news/local/michigan/2017/03/14/lawsuit-defective-robot-killed-factory-worker-human-error-blame/99173888/. Accessed 3 May 2023
3. Chen, T., Phan, L.T.X.: SafeMC: a system for the design and evaluation of mode-change protocols. In: Proceedings of the IEEE Real-Time and Embedded Technology and Applications Symposium, RTAS, pp. 105–116 (2018)
4. Dibaji, S.M., Pirani, M., Flamholz, D.B., Annaswamy, A.M., Johansson, K.H., Chakrabortty, A.: A systems and control perspective of CPS security. Annu. Rev. Control. **47**, 394–411 (2019)
5. Guiochet, J., Powell, D., Baudin, É., Blanquart, J.P.: Online safety monitoring using safety modes. In: Workshop on Technical Challenges for Dependable Robots in Human Environments, pp. 1–13 (2008)
6. Herkert, J., Borenstein, J., Miller, K.: The Boeing 737 MAX: lessons for engineering ethics. Sci. Eng. Ethics **26**(6), 2957–2974 (2020)
7. Inayat, I., Farooq, M., Inayat, Z., Abbas, M.: Security-based safety hazard analysis using FMEA: a DAM case study. In: Kotsis, G., et al. (eds.) DEXA 2021. CCIS, vol. 1479, pp. 18–30. Springer, Cham (2021). https://doi.org/10.1007/978-3-030-87101-7_3
8. Kim, S., Lee, J., Kang, C.: Analysis of industrial accidents causing through jamming or crushing accidental deaths in the manufacturing industry in South Korea: focus on non-routine work on machinery. Saf. Sci. **133**, 104998 (2021)
9. Lyu, X., Ding, Y., Yang, S.H.: Safety and security risk assessment in cyberphysical systems. IET Cyber-Phys. Syst. Theor. Appl. **4**(3), 221–232 (2019)
10. Malm, T., Salmi, T., Marstio, I., Montonen, J.: Dynamic safety system for collaboration of operators and industrial robots. Open Eng. **9**(1), 61–71 (2019)
11. Neukirchner, M., Quinton, S., Ernst, R., Lampka, K.: Multi-mode monitoring for mixed-criticality real-time systems. In: 2013 International Conference on Hardware/Software Codesign and System Synthesis, CODES+ISSS 2013 (2013)
12. Nikolakis, N., Maratos, V., Makris, S.: A cyber physical system (CPS) approach for safe human-robot collaboration in a shared workplace. Robot. Comput. Integr. Manuf. **56**, 233–243 (2019)
13. Niu, L., Sahabandu, D., Clark, A., Poovendran, R.: Verifying safety for resilient cyber-physical systems via reactive software restart. In: Proceedings of the 13th ACM/IEEE International Conference on Cyber-Physical Systems, ICCPS 2022, pp. 104–115 (2022)
14. Open Source Robotics Foundation: TurtleBot (2023). https://www.turtlebot.com/. Accessed 3 May 2023
15. Poltavtseva, M., Shelupanov, A., Bragin, D., Zegzhda, D., Alexandrova, E.: Key concepts of systemological approach to CPS adaptive information security monitoring. Symmetry **13**(12), 2425 (2021)
16. Riegler, M., Sametinger, J., Vierhauser, M.: A distributed MAPE-K framework for self-protective IoT devices. In: IEEE Proceedings of the 18th Symposium on Software Engineering for Adaptive and Self-Managing Systems, SEAMS 2023 (2023)
17. Vierhauser, M., Wohlrab, R., Stadler, M., Cleland-Huang, J.: AMon: a domain-specific language and framework for adaptive monitoring of cyber-physical systems. J. Syst. Softw. **195**, 111507 (2023)

18. Villani, V., Pini, F., Leali, F., Secchi, C.: Survey on human-robot collaboration in industrial settings: safety, intuitive interfaces and applications. Mechatronics **55**, 248–266 (2018)
19. Yin, H., Hansson, H.: Fighting CPS complexity by component-based software development of multi-mode systems. Designs **2**(4), 39 (2018)
20. Zacharaki, A., Kostavelis, I., Gasteratos, A., Dokas, I.: Safety bounds in human robot interaction: a survey. Saf. Sci. **127**, 104667 (2020)

An Intermediate Representation for Rewriting Cypher Queries

Daniel Hofer[1,2]([⊠]) [iD], Aya Mohamed[1,2] [iD], Stefan Nadschläger[1],
and Dagmar Auer[1,2] [iD]

[1] Institute for Application-oriented Knowledge Processing (FAW),
Johannes Kepler University (JKU) Linz, Linz, Austria
{daniel.hofer,aya.mohamed,stefan.nadschlaeger,dagmar.auer}@jku.at
[2] LIT Secure and Correct Systems Lab, Linz Institute of Technology (LIT),
Johannes Kepler University (JKU) Linz, Linz, Austria

Abstract. Some of the current graph database systems provide built-in authorization and access control features. However, many authorization requirements demand for more sophisticated access control such as fine-grained, attribute-based access control (ABAC). Therefore, we decided for a query rewriting approach to enforce these authorizations. We propose an intermediate representation for the semantics of the query. Based on the Cypher grammar, we build an abstract syntax tree (AST) of the query to be extended (i.e., rewritten). We consider a universal class hierarchy for our AST nodes based on the composite pattern, while the semantics of the nodes is introduced via data components. This provides flexibility with respect to the supported kinds of permissions and complexity of the Cypher queries. Our concept and prototypical implementation rely on ANTLR (ANother Tool for Language Recognition), which generates a parser based on the Cypher grammar to create and traverse concrete syntax trees.

Keywords: Abstract Syntax Tree (AST) · Query Rewriting · Cypher

1 Introduction

Consider we have a graph database using the Cypher query language such as *Neo4j* and others [6]. As not the whole data is public, we have to enforce access control. Some databases offer built-in authorization and access control. While the *community* edition of Neo4j does not provide access control, the *enterprise* edition supports role-based access control (RBAC) [5,8]. However, the authorization requirements in our current project demand attribute-based access control (ABAC). To enforce ABAC, we rely on query rewriting to filter the data according to the authorization requirements [4].

The initial challenge of rewriting Cypher queries to apply authorization-specific filters is identifying the semantics of the query. *OpenCypher* [6] provides a grammar for the Cypher query language. We build an abstract syntax tree (AST), which describes the semantics of the Cypher query to ease its rewriting

G. Kotsis et al. (Eds.): DEXA 2023 Workshops, CCIS 1872, pp. 86–90, 2023.
https://doi.org/10.1007/978-3-031-39689-2_9

for access control. However, the resulting AST might be complex. Hence, we seek to answer the following research questions:

RQ1 What are the semantic building blocks to define the AST for a Cypher query?
RQ2 What information must be encoded by the nodes of the AST?
RQ3 Can a prototypical implementation of the proposed concept be provided and applied to selected Cypher queries?

2   Related Work

To determine the appropriate structure of our AST, we need to consider the semantics necessary for query rewriting, but also implementation aspects. Van den Brand et al. [2] show the generation of a strongly typed AST in Java. They rely on immutable sub-trees and use reference equality for memory efficiency and reusability. While the strongly typed nodes would have an advantage on compile-time checks, we cannot make use of them as we are using ANTLR. It expects one common super-class and would hide specialized classes. Arusoaie et al. [1] use parser generators to automatically construct an AST for a context free grammar. They distinguish between *Parse Trees* and *Abstract Syntax Trees* in the sense that the latter focuses on the semantics and ignores information which is syntactic sugar only - a task we are about to do. We use ANTLR to generate our parse tree from which we construct our AST by traversing the parse tree and only generate AST nodes if we need them. Chris Clark [3] proposed a very simple model for ASTs using only one node type to build the overall tree. Claimed benefits are that an operation for one node is applicable to every other node. Inspired by this work, we try to keep the number of classes used for the AST low and use an *enumeration* for further specifying the AST nodes.

3   Abstract Syntax Tree for Cypher Queries

Cypher is the declarative query language for the graph database *Neo4j*. Its grammar is defined by the *openCypher* project [6]. Many of the syntactical details in the concrete syntax tree are not relevant to our query rewriting approach. Therefore, we build a simplified AST with focus on the relevant semantics of Cypher queries to add/modify filters and generate a textual query. Currently, we focus on the authorized reading of data (i.e. MATCH) and omit clauses such as CREATE, DELETE or WITH.

3.1   Types of AST Nodes

We start with the Cypher syntax for a node with a label, i.e., (:Label). The relevant part of the grammar in EBNF is shown in Listing 1. Many details of the concrete syntax tree from Listing 2 are not needed for query rewriting (e.g.,

```
1  NodePattern = '(', [SP], [Variable, [SP]], [NodeLabels, [SP]],
   ↪ [Properties, [SP]], ')' ;
2  NodeLabels = NodeLabel, { [SP], NodeLabel } ;
3  NodeLabel = ':', [SP], LabelName ;
4  LabelName = SchemaName ;
5  SchemaName = SymbolicName | ReservedWord ;
6  SymbolicName = UnescapedSymbolicName | EscapedSymbolicName | ... ;
```

Listing 1. EBNF rules required for parsing (:Label) (cut off at SymbolicName)

```
1  // Concrete syntax tree
2  NodePattern('(', NodeLabels(NodeLabel(
3     ':', LabelName(SchemaName(SymbolicName('Label')))))),')')
4
5  // Simplified abstract syntax tree
6  Node(Label('Label'))
```

Listing 2. Concrete syntax tree and simplified AST for Listing 1

parenthesis and colon). All we need to know is that we have a node with a label. This information is sufficient for processing the semantics and generating the textual representation. We can also formally define a Cypher *Node* for our AST with an optional *Variable(v)*, a set of labels L and a set of properties P. Both, L and P might be empty. v and l denote a string. Properties P require complex expressions and are (as in Listing 1) omitted due to space limitations:

$$Node = \{Variable(v)\} \cup L \cup P$$
$$L = \{Label(l_1), ..., Label(l_n)\}$$

3.2   AST Classes

For a prototype, we also need a data structure for our AST. To keep the number of classes small (cp. Clark [3]), we do not generate a class per AST type. Instead, we only distinguish between internal nodes and leaf nodes (with and without value). Therefore, we can implement our AST using four classes which implement the composite pattern [7]. The root of the class hierarchy is the abstract class *AstNode*. All three concrete classes, i.e., *AstInternalNode*, *AstLeafValue*, and *AstLeafNoValue* are directly derived from *AstNode* as shown in Fig. 1a.

All nodes in our AST contain a data component *type*, which specifies the corresponding part of the Cypher query. The concrete classes are:

- **AstInternalNode** is the root to a sub-AST describing one aspect of the query (e.g., a Cypher node).
- **AstLeafValue** is the leaf node type containing a value (e.g., a node's label).
- **AstLeafNoValue** describes parts of the query which have neither children nor a value (e.g., keywords like *DISTINCT*).

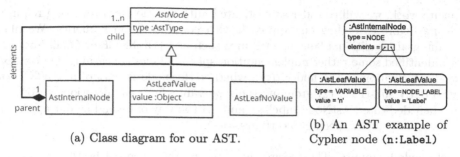

(a) Class diagram for our AST.

(b) An AST example of Cypher node (n:Label)

Fig. 1. Class diagram and an example of our AST.

4 Implementation

Our prototypical implementation[1] is based on Kotlin. We use ANTLR to build the Cypher parser and parse trees for the queries. From each parse tree, we extract the relevant information and store them in our AST (cp. Sect. 3.2). The hierarchical structure of the AST describes the structure of the query. The nodes of our AST have different classes depending on their role in the tree structure, i.e., internal node or leave (with or without value). The data components (e.g. the *type* property) represent the semantic information of the AST node. *AstLeafValue* nodes contain information, such as values for variables and labels (see Fig. 1b). In contrast to the parse tree, only the parts which are necessary for the query rewriting are added to the AST. Our approach is flexible as no new classes are required when adding new components.

5 Conclusion

In this paper, we proposed constructing an abstract syntax tree (AST) as an easy-to-use intermediate representation for rewriting Cypher queries in the context of access control. We used ANTLR and the Cypher grammar provided by the *openCypher* project to generate a Cypher parser that builds a parse tree for each Cypher query to be rewritten. Only the semantic information of the query (RQ1) and not the syntax details are stored in our AST. This is sufficient to modify the content of the query and to generate a textual representation. Our second contribution (RQ2) is a datamodel for our AST. We used the composite pattern to define a flexible data structure, which stores arbitrary nodes for our AST. All semantic information is specified via properties. Therefore, we can easily add new types without modifying the AST classes. This is especially useful in future work to support additional query types.

We implemented a prototype using Kotlin and ANTLR (RQ3). Our full version of this paper[2] contains more examples and details of the prototype. In

---

[1] https://github.com/jku-lit-scsl/CypherRewritingCore.

[2] https://research.daho.at/papers/ast_for_cypher_queries.

future work, we will consider sophisticated authorization and rewriting requirements before extending our approach with complex Cypher queries. We plan to integrate additional language features such as WITH, MERGE or CALL. Further, we identified some rather implementation-specific issues concerning: (1) contextual information in the nodes (e.g., whether the node is part of a MATCH or RETURN sub-tree), (2) grouping the *type* values into enumerations according to their dedicated AstNode subclass, and (3) the way of extending our approach (i.e., generic [9] or specific to certain types).

Acknowledgements. This research has been partly supported by the LIT Secure and Correct Systems Lab funded by the State of Upper Austria and by the COMET-K2 Center of the Linz Center of Mechatronics (LCM) funded by the Austrian federal government and the federal state of Upper Austria.

References

1. Arusoaie, A., Vicol, D.I.: Automating abstract syntax tree construction for context free grammars. In: 2012 14th International Symposium on Symbolic and Numeric Algorithms for Scientific Computing, pp. 152–159 (2012). https://doi.org/10.1109/SYNASC.2012.8
2. Van den Brand, M., Moreau, P.E., Vinju, J.: Generator of efficient strongly typed abstract syntax trees in java. IEE Proc.-Softw. **152**(2), 70–78 (2005)
3. Clark, C.: Uniform abstract syntax trees. ACM SIGPLAN Notices **35**(2), 11–16 (2000)
4. Hofer, D., Nadschläger, S., Mohamed, A., Küng, J.: Extending authorization capabilities of object relational/graph mappers by request manipulation. In: Strauss, C., Cuzzocrea, A., Kotsis, G., Tjoa, A.M., Khalil, I. (eds.) Database and Expert Systems Applications, pp. 71–83. Springer International Publishing, Cham (2022). https://doi.org/10.1007/978-3-031-12426-6_6
5. Mohamed, A., Auer, D., Hofer, D., Küng, J.: Authorization and access control for different database models: Requirements and current state of the art. In: Dang, T.K., Küng, J., Chung, T.M. (eds.) Future Data and Security Engineering. Big Data, Security and Privacy, Smart City and Industry 4.0 Applications, pp. 225–239. Springer Nature Singapore, Singapore (2022). https://doi.org/10.1007/978-981-19-8069-5_15
6. Neo4j Inc: openCypher. https://opencypher.org/. Accessed 13 Feb 2023
7. Riehle, D.: Composite design patterns. In: Proceedings of the 12th ACM SIGPLAN Conference on Object-oriented Programming, Systems, Languages, and Applications, pp. 218–228 (1997)
8. Sicari, S., Rizzardi, A., Coen-Porisini, A.: Security&privacy issues and challenges in NOSQL databases. Comput. Netw. **206**, 108828 (2022). https://doi.org/10.1016/j.comnet.2022.108828, https://www.sciencedirect.com/science/article/pii/S1389128622000470
9. Wadler, P., et al.: The expression problem. Posted on the Java Genericity mailing list (1998). https://homepages.inf.ed.ac.uk/wadler/papers/expression/expression.txt

An Effective Feature Selection for Diabetes Prediction

In-ae Kang[1], Soualihou Ngnamsie Njimbouom[1], and Jeong-Dong Kim[1,2,3]([⊠])

[1] Department of Computer Science and Electronic Engineering, Sun Moon University, Asan-Si 31460, Republic of Korea
{inae5004,salehrico,kjd4u}@sunmoon.ac.kr
[2] Division of Computer Science and Engineering, Sun Moon University, Asan-Si 31460, Republic of Korea
[3] Genome-Based BioIT Convergence Institute, Sun Moon University, Asan-Si 31460, Republic of Korea

Abstract. With the rapid advancement of technology and the ever increasing amount of data in the healthcare domain, big data analytics has become a significant study area. Analyzing patterns in patient treatment for the early detection and diagnosis of diseases can improve overall healthcare quality. Machine learning has emerged as a promising technology for aiding clinicians in making accurate diagnosis decisions. In this paper, we aim to propose an approach through the feature selection technique and employing various ML algorithms such as GBDT, NB, K-NN, SVM, LR, RF, and DT that will identify the subset of features relevant to the prediction of diabetes disease. The performance of each algorithm is evaluated using the Pima Indians Diabetes Dataset and Korean National Health and Nutrition Dataset. Experimental results show that the GBDT algorithm performs the best in predicting the disease with the highest accuracy.

Keywords: Decision support system · Disease Prediction · Feature selection

1 Introduction

Diabetes mellitus, a persistent metabolic illness characterized by hyperglycemia, is rising in prevalence due to socioeconomic development [1, 2]. Poor management of diabetes can lead to organ damage, mortality, and impaired physiological systems like ocular, cardiovascular, renal, neuropathic, and podiatric systems [3, 4]. Globally, the prevalence of diabetes was estimated at 451 million individuals in 2017, projected to rise to 693 million in the next 26 years [5]. The development of diabetes involves a combination of environmental and genetic factors.

Artificial Intelligence (AI) and Machine Learning (ML) Advancements have significantly enhanced automated illness detection and diagnosis [6, 7]. Computer-based systems have shown promise in efficient disease diagnosis and demonstrated superior accuracy and reliability compared to human capabilities. AI and ML methods have been widely used in building automated diabetes diagnosis systems, utilizing medical data

© The Author(s), under exclusive license to Springer Nature Switzerland AG 2023
G. Kotsis et al. (Eds.): DEXA 2023 Workshops, CCIS 1872, pp. 91–96, 2023.
https://doi.org/10.1007/978-3-031-39689-2_10

mining and knowledge exploration, therein helping diabetes patient stratification and treatment, as demonstrated in recent studies. [8–11].

However, current decision support systems for diabetes diagnosis encounter some limitations. The widely utilized Pima Indians Diabetes Dataset (PIDD), employed in ML studies, exhibits a relatively low accuracy of around 70% due to limited sample sizes and the exclusion of critical health-related variables such as insulin, obesity, and genetic factors. Additionally, the data collected during the 1980s may not adequately represent the present population or reflect current practices in diabetes diagnosis.

To address these limitations, the proposed study aims to leverage a new dataset derived from the comprehensive 2021 National Nutrition Survey for diabetes diagnosis. By training ML algorithms on this updated and relevant diabetes dataset, reliable accuracy can be achieved, facilitating the identification of predisposing factors associated with diabetes in the current population and using the most recent and relevant information.

2 Related Work

Previous research has utilized various ML algorithms for illness classification, diagnosis, prediction, and therapy [8–10, 12], yet they have yet to achieve an accuracy exceeding 80%. Various ML methods have been employed by researchers for diabetes detection; for instance, Chatrati et al. [13] used Support Vector Machine (SVM) algorithm for accurate hypertension and diabetes prediction based on key health parameters. Goyal et al. [14] and Prakash et al. [15] utilized ensemble techniques for early type 2 diabetes prediction achieving maximal accuracy of 77.60% and 79.22%, respectively. FA Khaleel et al. [11] used Logistic Regression (LR), Naïve Bayes (NB), and K-Nearest Neighbor (KNN) algorithms, with LR exhibiting higher efficiency and achieving remarkable accuracy of 94%.

In the Internet of Medical Things (IoMT) context, Victor Chang et al. [16] and Jackins et al. [17] leveraged the RF model showing its superior performance compared to other models in diabetes detection. Sneha et al. [18] utilized optimal feature selection methods, with NB yielding the best accuracy and RF exhibiting the highest specificity. Hasan et al. [19] employed correlation and Principal Component Analysis (PCA) for feature selection and ensemble classifiers, achieving the highest Area Under the Curve (AUC) using AdaBoost and Gradient ensembles. Saxena et al. [20] explored the performance of Multilayer Perceptron, Decision Tree (DT), KNN, and RF classifiers, with RF achieving the highest accuracy of 79.8%.

3 Proposed Method

The proposed diabetes diagnosis model, shown in Fig. 1, consists of three main steps: data collection, preprocessing, and prediction. Data collection steps consist of sourcing data from the PIDD (UCI ML Repository) and the 2021 Korean National Health and Nutrition Dataset (KNHND) from the CDC [21]. The preprocessing step employs Information Gain (IG) and Mutual Information (MI) techniques for feature selection. Multiple ML models, such as Gradient Boosting Decision Trees, KNN, Random Forest, DT, LR, SVM, and NB, are then used to accurately detect diabetes.

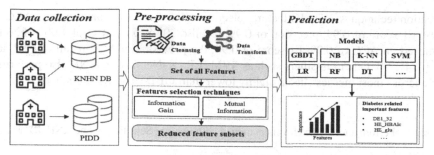

Fig. 1. Conceptual Model.

3.1 Data Collection

This study utilized the PIDD and the KNHND datasets. The PIDD includes 1,543 instances and 9 attributes, 530 instances represent diabetes (labeled as "1"), and 1,013 instances represent non-diabetes (labeled as "0"). On the other hand, the KNHND comprises health questionnaires and examinations from 5,239 respondents, with 593 instances (11%) of diabetic conditions and 4,646 instances (89%) without diabetic conditions. The KNHND has 42 features, one of which acts as a label: "0", denoting diabetes, and "1", denoting non-diabetes.

3.2 Pre-processing

Our preprocessing began with data cleaning, removing features with many empty cells; following that, we performed feature selection via IG and MI techniques, eliminating irrelevant features and retaining essential ones for optimal prediction [22]. IG, a filter-based method, gauges the predictor variable's categorizing ability regarding the dependent variable, leveraging information theory to compute the statistical dependence [23]. Conversely, MI quantifies the dependency between two variables, identifying both linear and non-linear correlations, thus frequently chosen for feature selection [24].

3.3 Prediction

Our prediction module consisted of experimenting with seven ML methods on the subset of features selected via IG and MI techniques. Each dataset was partitioned into training and test sets. The training set was utilized to train the models, utilizing 10-fold cross-validation for hyperparameter tuning, enhancing generalizability, and avoiding overfitting. The test set was used to evaluate the ML methods.

4 Experimentation

4.1 Result

Performance of the Classifiers with Feature Selection. Table 1 presents the performance of various ML classifiers trained on the KNHND dataset using IG and MI feature selection techniques. With IG feature selection, GBDT achieved an accuracy (ACC) of 0.99, F1-score of 0.99, precision of 0.99, recall of 0.98, and AUC of 1.00, while RF achieved an ACC of 0.99, F1-score of 0.98, the precision of 1.00, recall of 0.97, and AUC of 1.00. Other classifiers such as SVM, LR, K-NN, DT, and NB also performed well across most metrics, albeit slightly lower than GBDT and RF classifiers.

Table 1. Performance of the classifiers trained with Feature Selection on the KNHND dataset.

KNHND	IG					MI				
	ACC	F1-score	Precision	Recall	AUC	ACC	F1-score	Precision	Recall	AUC
GBDT	**0.99**	**0.99**	**0.99**	**0.98**	1.00	0.97	0.97	0.98	0.96	1.00
RF	0.99	0.98	1.00	0.97	1.00	**0.99**	**0.98**	**0.99**	**0.97**	**1.00**
SVM	0.99	0.99	0.98	0.99	1.00	0.98	0.98	0.99	0.97	1.00
LR	0.99	0.97	0.98	0.96	1.00	0.97	0.97	0.98	0.95	0.99
K-NN	0.97	0.97	0.95	0.98	1.00	0.96	0.96	0.94	0.97	1.00
DT	0.96	0.96	0.96	0.95	0.99	0.97	0.96	0.97	0.95	0.99
NB	0.95	0.94	0.95	0.93	0.99	0.96	0.95	0.96	0.93	1.00

Table 2 presents the performance of ML classifiers, trained with IG and MI feature selection techniques, on the PIDD dataset. Utilizing IG, RF achieves an ACC of 0.93, F1-score of 0.98, precision of 1.00, recall of 0.97, and AUC of 0.96. Other classifiers, such as GBDT, K-NN, and SVM, perform relatively well in accuracy using IG and MI feature selection techniques, albeit marginally lower than RF. These findings underline the significance of selecting appropriate feature selection techniques in enhancing ML models' predictive accuracy and generalizability.

Table 2. Performance of the classifiers trained with Feature Selection on the PIDD dataset.

PIDD	IG					MI				
	ACC	F1-score	Precision	Recall	AUC	ACC	F1-score	Precision	Recall	AUC
RF	**0.93**	**0.98**	**1.00**	**0.97**	**0.96**	**0.91**	**0.89**	**0.89**	**0.89**	**0.96**
GBDT	0.90	0.87	0.87	0.87	0.97	0.88	0.85	0.87	0.84	0.97
K-NN	0.90	0.87	0.86	0.89	0.89	0.89	0.86	0.87	0.85	0.95
SVM	0.87	0.83	0.84	0.82	0.95	0.87	0.83	0.85	0.82	0.95
DT	0.77	0.71	0.71	0.72	0.84	0.85	0.81	0.82	0.79	0.90
LR	0.76	0.69	0.71	0.66	0.85	0.77	0.69	0.75	0.63	0.85
NB	0.71	0.62	0.66	0.58	0.81	0.74	0.66	0.71	0.63	0.81

5   Conclusion

The results demonstrate that utilizing IG and MI for feature selection enhanced the performance of classifiers, notably GBDT and RF, in predicting diabetes incidence using the KNHND dataset. These classifiers outperformed others across multiple evaluation metrics, including ACC, F1-score, precision, recall, and AUC. Additionally, the study evaluated the performance of classifiers trained with feature selection on the PIDD, revealing that RF achieved the highest performance in most metrics. At the same time, GBDT, K-NN, and SVM also exhibited good results. These findings underline the value of feature selection techniques in developing accurate ML models for diabetes detection.

References

1. Sun, Y., Zhang, D.: Machine learning techniques for screening and diagnosis of diabetes: a survey. Teh. Vjesn. **26**, 872–880 (2019)
2. Ndisang, J.F., Vannacci, A., Rastogi, S.: Insulin resistance, type 1 and type 2 diabetes, and related complications 2017. J. Diabetes Res. **2017**, e1478294 (2017). [PubMed]
3. Malik, S., Harous, S., El-Sayed, H.: Comparative analysis of machine learning algorithms for early prediction of diabetes mellitus in women. In: Chikhi, S., Amine, A., Chaoui, A., Saidouni, D.E., Kholladi, M.K. (eds.) MISC 2020. LNNS, vol. 156, pp. 95–106. Springer, Cham (2021). https://doi.org/10.1007/978-3-030-58861-8_7
4. Himsworth, H.P., Kerr, R.B.: Insulin-sensitive and insulin-insensitive types of diabetes mellitus. Clin. Sci. **4**, 119–152 (1939)
5. World Health Organization, 2020 World Health Organization. https://www.who.int/news-room/fact-sheets/detail/diabetes. Accessed 8 June 2020
6. Theera-Umpon, N., Poonkasem, I., Auephanwiriyakul, S., Patikulsila, D.: Hard exudate detection in retinal fundus images using supervised learning. Neural Comput. Appl. **32**(17), 13079–13096 (2019). https://doi.org/10.1007/s00521-019-04402-7
7. Afzali, S., Yildiz, O.: An effective sample preparation method for diabetes prediction. Int. Arab J. Inf. Technol. **15**(6), 968–973 (2018)
8. Jaiswal, V., Negi, A., Pal, T.: A review on current advances in machine learning based diabetes prediction. Prim. Care Diabetes **15**, 435–443 (2021)
9. Tariq, H., Rashid, M., Javed, A., Zafar, E., Alotaibi, S.S., Zia, M.Y.I.: Performance analysis of deep-neural-network-based automatic diagnosis of diabetic retinopathy. Sensors **22**, 205 (2022)
10. Kumar, D., et al.: Automatic detection of white blood cancer from bone marrow microscopic images using convolutional neural networks. IEEE Access **8**, 142521–142531 (2020)
11. Khaleel, F.A., Al-Bakry, A.M.:Diagnosis of diabetes using machine learning algorithms. Mater. Today: Proc. (2021)
12. Saxena, R., Sharma, S.K., Gupta, M., Sampada, G.C.: A comprehensive review of various diabetic prediction models: a literature survey. J. Healthc. Eng. **2022**, e8100697 (2022). [PubMed]
13. Chatrati, S.P., et al.: Smart home health monitoring system for predicting type 2 diabetes and hypertension. J. King Saud Univ.—Comput. Inf. Sci. **34**, 862–870 (2020)
14. Goyal, P., Jain, S.: Prediction of type-2 diabetes using classification and ensemble method approach. In: Proceedings of the 2022 International Mobile and Embedded Technology Conference (MECON), Noida, India, pp. 658–665, 10–11 March 2022
15. Prakash, A.: An ensemble technique for early prediction of type 2 diabetes mellitus—a normalization approach. Turk. J. Comput. Math. Educ. **12**, 9 (2021)

16. Chang, V., et al.: Pima Indians diabetes mellitus classification based on machine learning (ML) algorithms.Neural Comput. Appl., 1-17 (2022)
17. Jackins, V., Vimal, S., Kaliappan, M., Lee, M.Y.: AI-based smart prediction of clinical disease using random forest classifier and Naive Bayes. J. Supercomput. **77**(5), 5198–5219 (2020). https://doi.org/10.1007/s11227-020-03481-x
18. Sneha, N., Tarun, G.: Analysis of diabetes mellitus for early prediction using optimal feature selection. J. Big data **6**, 3 (2019)
19. Kamrul Hasan, M., Ashraful Alam, M., Das, D., Hussain, E., Hasan, M.: Diabetes prediction using ensembling of different machine learning classifiers.IEEE Access **8** (2020). Article ID: 76531
20. Saxena, R., Sharma, S.K., Gupta, M., Sampada, G.C.: A novel approach for feature selection and classification of diabetes mellitus: machine learning methods. Comput. Intell. Neurosci. **2022**, e3820360 (2022)
21. Korea Centers for Disease Control and Prevention. https://knhanes.kdca.go.kr/knhanes/sub03/sub03_02_05.do
22. Khaire, U.M., Dhanalakshmi, R.: Stability of feature selection algorithm: a review. J. King Saud Univ. Comput. Inf. Sci. (2019)
23. Gao, Z., Xu, Y., Meng, F., Qi, F., Lin, Z.: Improved information gain-based feature selection for text categorization. In: Proceedings of the 2014 4th International Conference on Wireless Communications, Vehicular Technology, Information Theory and Aerospace Electronic Systems (VITAE), IEEE, Aalborg, Denmark, pp. 1–5, 11–14 May 2014
24. Li, J., et al.: Feature selection: a data perspective. ACM Comput. Surv. **50**, 1–45 (2017)

Author Index

G. Kotsis et al. (Eds.): DEXA 2023 Workshops, CCIS 1872, p. 97, 2023.
https://doi.org/10.1007/978-3-031-39689-2

Printed in the United States
by Baker & Taylor Publisher Services